LEVERAGING BIOMEDICAL AND HEALTHCARE DATA

LEVERAGING BIOMEDICAL AND HEALTHCARE DATA

Semantics, Analytics and Knowledge

Edited by

FIRAS KOBEISSY

ALI ALAWIEH

FADI A. ZARAKET

KEVIN WANG

ACADEMIC PRESS

An imprint of Elsevier

Academic Press is an imprint of Elsevier
125 London Wall, London EC2Y 5AS, United Kingdom
525 B Street, Suite 1650, San Diego, CA 92101, United States
50 Hampshire Street, 5th Floor, Cambridge, MA 02139, United States
The Boulevard, Langford Lane, Kidlington, Oxford OX5 1GB, United Kingdom

Notices
Knowledge and best practice in this field are constantly changing. As new research and experience
broaden our understanding, changes in research methods, professional practices, or medical treatment
may become necessary.

Practitioners and researchers must always rely on their own experience and knowledge in evaluating
and using any information, methods, compounds, or experiments described herein. In using such
information or methods they should be mindful of their own safety and the safety of others, including
parties for whom they have a professional responsibility.

To the fullest extent of the law, neither the Publisher nor the authors, contributors, or editors, assume
any liability for any injury and/or damage to persons or property as a matter of products liability,
negligence or otherwise, or from any use or operation of any methods, products, instructions, or ideas
contained in the material herein.

Library of Congress Cataloging-in-Publication Data
A catalog record for this book is available from the Library of Congress

British Library Cataloguing-in-Publication Data
A catalogue record for this book is available from the British Library

ISBN 978-0-12-809556-0

For information on all Academic Press publications visit our
website at https://www.elsevier.com/books-and-journals

Working together
to grow libraries in
developing countries

www.elsevier.com • www.bookaid.org

Publisher: Stacy Masucci
Acquisition Editor: Rafael Teixeira
Editorial Project Manager: Mariana Kuhl
Production Project Manager: Sreejith Viswanathan
Cover Designer: Victoria Pearson

Typeset by SPi Global, India

DEDICATION

The authors, contributors, and editors would like to dedicate this modest scientific contribution to their families, parents, and significant others for their continued support, love, and patience.

CONTENTS

CONTRIBUTORS

Denes V. Agoston
Department of Anatomy, Physiology and Genetics, Uniformed Services University, Bethesda, MD, United States

Alanoud Al Jaberi
College of Information Technology, United Arab Emirates University, Al Ain, United Arab Emirates

Ali Alawieh
Department of Microbiology and Immunology, Medical University of South Carolina, Charleston, SC, United States; Department of Electrical and Computer Engineering, American University of Beirut, Beirut, Lebanon

L. Angelis
Department of Informatics, Aristotle University of Thessaloniki, Thessaloniki, Greece

Hiba Arnaout
Department of Computer Science, American University of Beirut, Beirut, Lebanon

Ehsaneddin Asgari
Molecular Cell Biomechanics Laboratory, Department of Bioengineering and Department of Mechanical Engineering, University of California, Berkeley, Berkeley, CA, United States

Javier Carnicero
Health Service of Navarra, Pamplona, Spain

Deepak Chhabra
Optimization and Mechatronics Laboratory, Department of Mechanical Engineering, University Institute of Engineering and Technology, Maharshi Dayanand University, Rohtak, India

Shady Elbassuoni
Department of Computer Science, American University of Beirut, Beirut, Lebanon

Wassim El-Hajj
Department of Computer Science, American University of Beirut, Beirut, Lebanon

Bilal Fadlallah
Department of Biomedical Engineering at Georgia Tech and Emory, Georgia Institute of Technology, Atlanta, GA, United States

Ali Fadlallah
Northern General Hospital, Sheffield, United Kingdom

Nabil Karnib
Department of Biochemistry and Molecular Genetics, American University of Beirut, Beirut, Lebanon

Firas Kobeissy
Department of Biochemistry and Molecular Genetics, American University of Beirut, Beirut, Lebanon; Program for Neurotrauma, Neuroproteomics & Biomarkers Research, Department of Emergency Medicine, University of Florida, Gainesville, FL, United States

K.A. Kyritsis
Laboratory of Pharmacology, School of Pharmacy, Aristotle University of Thessaloniki, Thessaloniki, Greece

Frédéric Mazurier
University of Tours, CNRS, GICC UMR 7292, LNOx Team, Tours, France

Hassan Pezeshgi Modarres
Departments of Bioengineering and Mechanical Engineering, Molecular Cell Biomechanics Laboratory, University of California Berkeley, Berkeley, CA, United States

Mohammad R.K. Mofrad
Molecular Cell Biomechanics Laboratory, Departments of Bioengineering and Mechanical Engineering, University of California Berkeley; Physical Biosciences Division, Lawrence Berkeley National Lab, Berkeley, CA, United States

Ali Nehme
University of Tours, CNRS, GICC UMR 7292, LNOx Team, Tours, France

Christos Ouzounis
Biological Computation & Process Laboratory, Chemical Process & Energy Resources Institute, Centre for Research & Technology Hellas, Thermi, Greece

Natalia Polouliakh
Sony Computer Science Laboratories Inc., Fundamental Research Laboratory; Systems Biology Institute, Tokyo; Department of Ophthalmology and Visual Science, Yokohama City University Graduate School of Medicine, Yokohama, Japan

Mahdi Razafsha
Department of Psychiatry, McKnight Brain Institute, University of Florida, Gainesville, FL, United States

David Rojas
Sistemas Avanzados de Tecnología (SATEC), Madrid, Spain

Zahraa Sabra
Department of Neurosciences, Medical University of South Carolina, Charleston, SC, United States; Department of Electrical and Computer Engineering, American University of Beirut, Beirut, Lebanon

Mohamed Sabra
Department of Microbiology and Immunology, Medical University of South Carolina, Charleston, SC, United States; Department of Electrical and Computer Engineering, American University of Beirut, Beirut, Lebanon

Pratyoosh Shukla
Enzyme Technology and Protein Bioinformatics Laboratory, Department of Microbiology, Maharshi Dayanand University, Rohtak, India

Chandana Tennakoon
College of Information Technology, United Arab Emirates University, Al Ain, United Arab Emirates

Rajat Vashistha
Optimization and Mechatronics Laboratory, Department of Mechanical Engineering, University Institute of Engineering and Technology, Maharshi Dayanand University, Rohtak, India

Ioannis Vizirianakis
Laboratory of Pharmacology, School of Pharmacy, Aristotle University of Thessaloniki, Thessaloniki, Greece

Kevin Wang
Program for Neurotrauma, Neuroproteomics & Biomarkers Research, Department of Emergency Medicine, University of Florida, Gainesville, FL, United States

Dinesh Yadav
Optimization and Mechatronics Laboratory, Department of Mechanical Engineering, University Institute of Engineering and Technology, Maharshi Dayanand University, Rohtak, India

Nazar Zaki
College of Information Technology, United Arab Emirates University, Al Ain, United Arab Emirates

Fadi A. Zaraket
Department of Electrical and Computer Engineering, Maroun Semaan faculty of engineering and architecture, American University of Beirut, Beirut, Lebanon

Kazem Zibara
ER045, PRASE, DSST; Biology Department, Faculty of Sciences-I, Lebanese University, Beirut, Lebanon

FOREWORD I

I am truly honored to introduce this book on "Leveraging Biomedical and Healthcare Data: Semantics, Analytics, and Knowledge." Since my early days as an undergraduate researcher in bioanalytical chemistry, I've thrived on the steep learning curve of scientific discovery. What could be more fun than studying something you are passionate about and leveraging the knowledge you gain to discover something new? There is always more to learn and more to do.

My research has focused on developing and applying novel proteomics methods and algorithms. First, I've discovered endogenous neuropeptides, biomarkers of traumatic brain injury, and biomarkers of multiple sclerosis. Second, I've discovered chemical and posttranslational modifications of therapeutic proteins such as monoclonal antibodies and antibody drug conjugates. Third, I've helped to get regulatory approval for these biomarkers and drugs for clinical trials. What work could provide me with more purpose than leveraging scientific discovery to improve the lives of patients and their families?

Systems biology tools are a natural fit for teams of interdisciplinary scientists exploring the bloom of information in the digital age. Throughout my career, I've benefited by sharing my proteomics and bioanalytical chemistry expertise with experts in other fields of inquiry, including neuroscience, neurosurgery, neuroimmunology, statistics, computer science, biomarkers, and drug development. For example, 14 years ago, I worked with Drs. Firas Kobeissy, Kevin Wang, and others on traumatic brain injuries. Proteomics enabled us to rapidly discover putative protein biomarkers from the thousands of proteins in brain tissue, based on the relative abundance of proteins in control versus test subjects. The strong biological insights of the team, particularly the role of calcium-dependent enzymes in generating endogenous peptides, broadened and deepened my own thinking to bring expected and unexpected synergies, including the first FDA approval of a blood test for a clinical brain concussion insult earlier this year.

This book is a terrific field guide for explorers of the scientific frontier of big data, from the beginner to the experienced adventurer. The uncharted territory is vast and ever expanding with no end in sight, so choose your team and systems biology tools wisely. Lastly, I encourage you to explore the interface: where courage matters most.

Most sincerely,
Will

William E. Haskins, Co–Founder and Chief Executive Officer
Gryphon Bio, Inc., South San Francisco, CA, United States

FOREWORD II

Knowledge understanding is a straightforward task when data have a unique meaning and are easily interpreted by automated digitization. However, in real life, this is not the case, as data are generated in an unclear semantics that requires further analysis and abstraction, especially with the neoform of data represented by the "Omics" field. In order to understand the true meaning of omics data, several curation steps need to be processed toward clarifying the real phenomena and interpretation of the given data. Those steps are required to represent, store, and reason the knowledge generated from such data. The causes of diseases or even their characterization are still unknown to many. Such knowledge is still unmatured and requires researchers to study and analyze it with different high-throughput platforms. On the other hand, the syntax of omics data needs to be processable and should have a clear structure; however, such data can even have noise and the structure is not yet clearly understood by scientists. We can say that until now, handling omics data is still a hot field for researchers. This book starts to handle such problems and gives some basic direction in the research of omics data. The distinction between bioinformatics and biomedicine needs to be clearly defined; this book provides a cutting-edge distinction between both areas of research.

The problem of understanding omics data is not only in the complexity of the syntax and semantics, but in the high throughput nature of such data, making rendering it unmanageable in several aspects. A best example of such complexity is the DNA data that have turned out to be extremely complex, requiring big data analytics for interpretation. Big data requires special computational mechanisms to process in terms of infrastructure and algorithms; a tractable solution has to be found. Big data scientists need to understand the nature of the problem in order to find the right algorithm for the machine to learn the knowledge behind it. Consequently, a book that explains the medical phenomena for computational scientists is crucial and reduces the gap between both research fields. As a matter of fact, the volume of data is not the only factor that requires special machine-learning algorithms, as the nature of the data also has such requirements in several aspects. First, the velocity of having such data sometimes requires processing in real time. Time-critical processing requires algorithms that process quickly without losing accuracy. Such problems are handled as a multiobjective task that has contradicting objectives. All these need to be considered, requiring optimization functions for their processing and interpretation. Second, the variability of data can add another dimension to the complexity of handling such types of data that requires the researchers' attention. Furthermore, omics data exist in different forms, structured and unstructured, whereas each requires different mechanisms and algorithms for handling. Both types of data will provide information that needs to be integrated to extract the right knowledge. For example, handling textual data is completely different than handling relational data, structured or unstructured. Textual data requires a natural language processing mechanism to translate the blocks of text into specific usable knowledge.

Networks are another mean of knowledge representation that can make a huge difference in transferring data-knowledge understanding between different researchers. This book provides a complete section for this topic. As the publications in the medical field exploded, the use of that literature needs to be defined in a useful manner; actually annotating knowledge of medical literature is a real path to medical evolution. Most of the medical directions try to integrate and link the literature given in that area in order for physicians to modify and standardize their practice. This required a lot of natural language processing mechanisms and analytics of publications. The phases of the literature collection are discussed in the first section of this book, *Understanding Molecular Architecture of Disease Using Big Data*.

Third, the veracity of the data generates a major problem for researchers. Data that are noisy, incomplete, or incorrect cannot be processed directly for a correct decision. Uncertain data have to be curated and verified, if not pruned to reduce the irrelevant or redundant data to reduce the complexity of such a problem. Much patient data is incomplete if not inaccurate, and that needs to be rectified while the accuracy of processing for a proper interpretation is another dimension that requires researchers' attention. Finally, such critical questions of accuracy and completeness need higher precision in analytics, which necessitates establishing novel methodologies and algorithms for handling such obstacles in data interpretation. In fact, there exist rare publications, specifically books that cover this specialization in bioinformatics. This particular topic is discussed fully in the second section of this book, *Guiding Health Care Decision Using Big Data*.

All these aforementioned dimensions need to be considered while analyzing the data to reach the actual value behind that disease-patient ecosystem. In order to generate the right knowledge, we need to analyze such complex data using effective complicated techniques, which introduces us to the area of precision medicine. The interdisciplinary nature of precision medicine requires that data scientists understand human biology and disease evolution; in parallel, physicians need to understand what knowledge can be generated using computational analysis. Reducing such a gap provides the fundamental basis for studying the human system and, hence, understanding disease causes and progression. Developing a vibrant understanding of the human body and disease structure in a patient leads us to define visual tools of special cases and, thus, to personalize the treatment. Personalized medicine is the current trend in medical research and methodology. The deeper our knowledge of personalized medicine, the better our understanding of the communities where such patients survive and interact. From that, researchers can then deduce generalized phenomena about such diseases. At that stage, we can precisely define causes and treatments of disease for those communities and associated diseases. Prediction of diseases is easily done through analysis of DNA and other omics data from birth.

To conclude, this book not only covers the basics of scientific understanding of medical phenomena but also reveals tools and datasets that need to be used in in silico research for these aims. The third section of this book, *Online Repositories and In-Silico Research in the Era of Big Data*, discusses the available datasets needed for bioinformatics research. Machine learning algorithms require a huge amount of historical data to develop models of prediction and association. The book actually provides a complete and divergent background for medical data analytics, which is the first step in the complex field of computational health informatics that definitely will lead to the maturity of precision medicine.

Passent Elkafrawy, Computer Science Professor
Faculty of Science, Department of Mathematics and Computer Science,
Menoufia University, Shebin ElKom, Egypt

PREFACE

The *systems biology discipline* emerged in the early 2000s as an integrative and holistic approach, aiming to assimilate knowledge through relatively large and complex amounts of data generated by high-throughput biological applications and tools. This amount of data led to an increase in novel findings and subsequently novel hypotheses. Yet the massive amount of biomedical research-related data introduced an ongoing data organization challenging task. This is exacerbated by the astronomical increase in the amount of data generated over mainly the last two decades, facilitated by the increased availability of novel high-throughput screening techniques such as whole genome sequencing, proteomics, and next-generation sequencing (NGS). The coupling of these techniques with relatively every field in biomedical research led to an exponential increase in the amount of data. Bioinformatics tools were created subsequently to apply automated reasoning to help analyze this data generated on a daily basis. The aim is to draw conclusions and translate them into more comprehensible findings that can be applied in the field of healthcare. In parallel, there was a zealous interest throughout the last decade in *semantic systems biology*: (1) integrating and sharing knowledge, (2) checking the quality of data, and (3) creating novel ways for computational support.

Although different "omics" aspects of systems biology such as next-generation genomics and proteomics have been extensively reviewed in the current literature, there is still limited availability of discussions devoted solely to semantic systems biology. To the best of our knowledge, there have been no previous books or extensive reports that discuss approaches used in the curation and analysis of text from published data, biomedical resources, and repositories as well as from medical records. There are also limited surveys of the future expectations of these approaches challenged by data dissemination and interpretation, resulting in massive scientific dataset accumulation. This book provides an overview of the novel approaches used in the mining and curation of big datasets, leading ultimately to data processing into disseminated knowledge. This book (i) provides an overview of the approaches used in semantic systems biology, (ii) introduces novel areas of application of semantic systems biology, and (iii) describes step-wise protocols for transforming heterogeneous data into useful knowledge that can influence healthcare and biomedical research. The book also reviews the main limitations and significant challenges that face such applications.

In this work, we assimilated a group of experts in both computational and translational health-related research with extensive expertise in omics fields. The leaders in this book infer from their expertise on the clinical and basic science levels direct applications of the herein discussed tools and applications. This book is divided into three parts.

The first part, "Understanding Molecular Architecture of Disease Using Big Data," introduces the high need for an efficient semantic system for analysis due to the presence of an immense amount of data from multiple approach "omics." The authors discuss the most efficient tools for the incorporation of large transcriptomics as well as proteomics data. The authors of the last two chapters of this part emphasize the high relevance of such approaches in several diseases, including traumatic brain injury (TBI) and ribosomopathies. The second part, "Guiding Health Care Decision Using Big Data," dissects the potent role of automated systems in guiding health care decisions. The authors in this part tackle the dependency of diagnosis and novel therapeutic strategies for neurodegenerative diseases on artificial intelligence, especially algorithms. The authors tackle epilepsy, particularly the diagnosis and predictability of seizures by automated systems. The discussion in this part moves to healthcare decision-making by relying on properly inferred knowledge from managed big amounts of data in medical records. A discussion of the computational modeling of a healthcare relevant topic, infectious disease surveillance, concludes this part. The third part, "Online Repositories and *In-Silico* Research in the Era of Big Data," focuses on the importance of using intuitive, friendly, and efficient annotation tools to navigate and edit annotated corpora relative to the present annotations. Authors in this part dedicate a section, explaining in a step-wise manner, an advanced annotation tool that can be applied in medical applications, taking the example of spinal cord injury (SCI) and stroke. The authors in this section conclude by focusing on the importance of finding proper data representation that is interpretable by machines. They introduce a novel language model-based distributed representation method for biological sequences and discuss the challenges associated with adopting such a language.

Finally, this book would not be possible without the cooperation of a large group of experts in this field. Each author added their special expertise, making this book a rich yet easily readable book with an immense amount of applications that can help every reader in a multitude of models. We hope that the readers will find it easy to apply what is provided in this book, leading to growth and evolution as a scientific community in an age where data is amply available.

Firas Kobeissy
Fadi A. Zaraket
Ali Alawieh
Kevin Wang

ACKNOWLEDGMENTS

First, we would like to send a great thank you to everyone who contributed to this book, including the authors, editors, and colleagues who all invested their time in this project. The great amount of cooperation between the authors and editors was a blessing, making the writing of this book an enjoyable journey. We also extend our gratitude to the authors for delivering years of their expertise and work in different areas of bioinformatics into this book in order to reach a consensus in this dazzling world of data. The herein discussed topics and applications are of great value and offer high-value knowledge in the areas of data mining, computational modeling, bioinformatics, and finally big data fields that can be applied in several health-related models. Furthermore, we would like to thank Nabil Karnib for his help in editing and chapter compilation.

Cover Art Design:
Designed by **Dr. Hisham F. Bahmad, MD, MSc**
Research Fellow
Department of Anatomy, Cell Biology, and Physiological Sciences
Faculty of Medicine
American University of Beirut
Beirut, Lebanon

CHAPTER 1

Comprehensive Workflow for Integrative Transcriptomics Meta-Analysis

Ali Nehme*, Frédéric Mazurier*, Kazem Zibara[†,‡]
*University of Tours, CNRS, GICC UMR 7292, LNOx Team, Tours, France
[†]ER045, PRASE, DSST, Lebanese University, Beirut, Lebanon
[‡]Biology Department, Faculty of Sciences-I, Lebanese University, Beirut, Lebanon

1. INTRODUCTION

Large-scale data analysis has become the first-line tool in medical and molecular biology, especially with the advancement in transcriptomics and RNAseq technologies, which provide important information on both gene expression and mutations. In fact, with the emergence of systems biology, the new "hypothesis-neutral" design of studies is replacing the traditional "hypothesis-driven" approach. Indeed, with the advancements in biostatistics and machine learning techniques during the last decade, transcriptomic analysis has been extensively used in hypothesis generation and validation. Therefore, the exponentially growing large-scale data require the construction of accessible data repositories.

Various databases are now available that offer the ability to download large-scale data from previous studies and provide the opportunity to reanalyze data in a cost- and time-efficient manner. However, this also gives the possibility for horizontal integration of data from multiple studies, which: (1) offers a higher level of statistical power with larger number of samples, (2) offers heterogeneity in available samples with more studied conditions and heterogeneous population, and (3) eliminates the effect of experimental bias during manipulation and sample preparation. In this chapter, we provide a comprehensive workflow for horizontal transcriptomics data integration. The workflow will focus on Affymetrix microarrays data due to their wide use in medical and molecular biology fields.

2. DATA PREPARATION

The experimental workflow for data integration consists of 4 main steps: data download, quality control, preprocessing, and data analysis (Fig. 1). Each of these steps includes different tasks that are described in detail in the following sections.

Leveraging Biomedical and Healthcare Data
https://doi.org/10.1016/B978-0-12-809556-0.00001-0

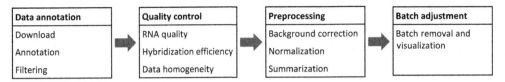

Fig. 1 Workflow chart of horizontal data integration. The workflow for data integration consists of four main steps: data download, quality control, preprocessing, and data analysis. After data download and annotation, samples are filtered based on certain inclusion and exclusion criteria. The latter includes, in addition to biological annotations, quality criteria that assess RNA integrity and samples homogeneity. The retained samples are then preprocessed, normalized, and batch-adjusted for further analysis.

2.1 Data Download

The first step is to download, annotate, and filter data based on previously defined criteria. Several public databases are available in accessible and comprehensive data repositories that are regularly updated. Of importance, most transcriptomic data are available on the Gene Expression Omnibus (GEO) database (Barrett et al., 2013). In addition, a large part of these data is also available on the ArrayExpress database from the European Bioinformatics Institute (EMBL-EBI) (Parkinson et al., 2009). However, specialized databases exist for specific diseases when the researcher is interested in vertical data integration. For example, The Cancer Genome Atlas (TCGA) comprises genomics, transcriptomics, chipseq, and proteomics data for different types of cancers (Tomczak et al., 2015). On the other hand, centralized public data repositories for proteomics data include the PRoteomics IDEntifications (PRIDE) database (Vizcaíno et al., 2016) and Proteomicsdb (Wilhelm et al., 2014), while the Human Metabolome Database (HMDB) is a repository for metabolomics data (Wishart et al., 2013).

In order to find relevant studies, one can query the search engine using specific keywords pertinent for their study of interest. In addition, several filters are available that allow the researcher to reduce the number of results and easily find specific studies. For example, the GEO database offers filters for organism (i.e., human, mouse, rat), study type (i.e., expression profiling, sequencing, genome binding, methylation), attributes (tissue or strain), sample count, and others. A link for each study will take the researcher to a new page that contains information on the study design and sample characteristics. Using this page, one can download sample annotation files, normalized data (series matrix), and raw data (CEL files). In fact, data can be directly downloaded through the R software using specific packages from the Bioconductor project (Huber et al., 2015), such as the *GEOquery* (Davis and Meltzer, 2007), *ArrayExpress* (Kauffmann et al., 2009a), and *TCGAbiolinks* (Colaprico et al., 2016) packages (see Table 1).

One of the critical steps in the workflow is data annotation, which provides the basis for samples inclusion for further analysis (Tseng et al., 2012; Hamid et al., 2009). In the GEO database, sample annotations are usually present in the series matrix files, while in

Table 1 R/Bioconductor packages used in horizontal data integration

Step	R/Bioconductor packages
Download	GEOquery, ArrayExpress, TCGAbiolinks
RNA quality control	simpleaffy, affyQCReport, arrayQualityMetrics
Data homogeneity	affyPLM, affycomp
Preprocessing	affy, gcrma, beadarray, lumi
Batch removal	sva, pamr, frma, limma, Harman, batchqc

the ArrayExpress database the annotation files are available as separate text files. Important annotations to include in any study comprise the species, tissue source, sample type (cell type), phenotype (normal, diseased, early stage, advanced stage), genotype, treatment, and survival information. One may include annotations that are only pertinent for the analysis. However, it is highly advisable to include all available annotations so that more information is available for inclusion/exclusion criteria. For example, if an analysis is to be performed to detect differentially expressed genes between different genotypes of a disease, it is important to note the tissue of origin, if the cells were sorted or tissue biopsies were used, and if the cells were amplified in vitro or directly used in RNA measurements. This is highly critical to select samples with the highest homogeneity of RNA expression and exclude any variables that may affect the analysis. In addition, this may save time if another analysis using different criteria is desired. It is worth noting that study inclusion/exclusion could also be performed based on the number of samples. For example, one can exclude all studies with sample count <20. This will help in reducing unbalanced batch effect and increasing the homogeneity of data during analysis.

2.2 Quality Control

The most crucial step in data preparation for data integration is making sure that the samples are of good quality and homogeneous to be comparable (Wilson and Miller, 2005; Warren and Liu, 2004). Indeed, poor-quality arrays significantly affect subsequent analyses and, consequently, invalidate the interpretation.

2.2.1 Single-Array Metrics

Microarray preparation requires multiple steps including RNA extraction, cDNA synthesis, labeling, hybridization, and data reading. For this, Affymetrix and other microarray technologies have set several methods to perform quality control at each step. Because β-actin and GAPDH are expressed in most cell types and are relatively long genes that are represented by multiple probes, RNA quality is usually assessed by comparing the ratio of intensity of probes on the 5' side to those on the 3' side (Fig. 2). Alternatively, the RNA degradation plot proposes to plot the average intensity of each probe across all probe sets, ordered from the 5' to the 3' end (Wilson and Miller, 2005). Since RNA degradation starts from the 5' end of the molecule, we would expect probe intensities to be globally

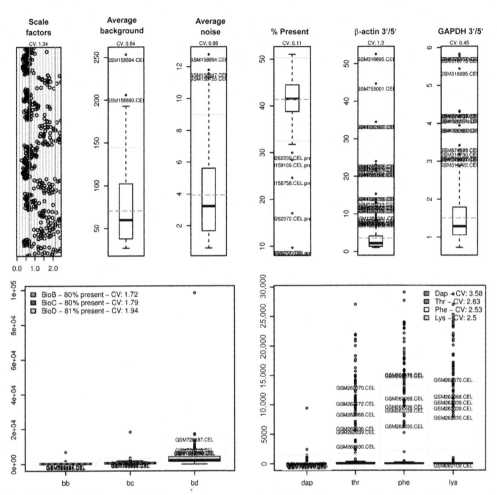

Fig. 2 Major criteria used in RNA quality assessment of affymetrix gene chips. Affymetrix have set several methods to perform quality control at each step of sample preparation and data generation. First samples' homogeneity and integrity can be assessed by scaling factor of normalization. Second, the average background and percentage of present calls evaluate global hybridization efficiency and RNA integrity in the samples. Third, RNA degradation can be weighed by the relative expression of 5′ to 3′ probes of GAPDH and actin. Finally, hybridization and cDNA synthesis efficiency are assessed using special spike-ins that are measured by specific probes in the microarray.

lowered at that end. Therefore, Affymetrix recommends that the 3′/5′ ratio should not exceed 3 for β-actin, and 1.25 for GAPDH (Fig. 2). A typical microarray experiment involves the hybridization of a cDNA, synthesized from RNA samples, to the DNA template probes. Assessment of cDNA synthesis efficiency is performed using sample preparation controls, which are spiked at the beginning of the chip process and used to assess the overall success of the target preparation steps. To verify that no bias occurred during the reverse transcription step between highly expressed and low expressed genes, Affymetrix designed spike-ins, Dap, Thr, Phe and Lys, which should be called present at a decreasing intensity (Dap > Thr > Phe > Lys) (Fig. 2). Finally, to evaluate cDNA labeling and hybridization efficiency, other spike-ins are used with different concentrations before the labeling step. These include the bioB, bioC, and bioD spike-ins, which should be called present in all microarrays (Fig. 2).

2.2.2 Multiarray Metrics

After making sure that all retained samples have a good RNA quality, data homogeneity is assessed to detect outliers. Several methods have been created for this purpose (Fig. 3). These methods usually rely on data after normalization to make sure that the detected outliers are shifted from the average fit of the microarrays. Boxplots and histograms of probe set intensity levels across microarrays are usually the first-line methods used to assess data homogeneity (Fig. 3). These plots show the distribution of probe sets intensities in each microarray. Integrated samples should show similar distribution, with similar median and quartiles. Similarly, data homogeneity can be assessed through heat maps of pairwise correlations between arrays. Pairwise correlations should be high since arrays are supposed to have similar distribution of their probe set intensities; however, outlier datasets should show lower correlations with the bulk of datasets (Kauffmann et al., 2009b).

Professionals in microarray analysis and quality metrics suggested new methods to assess microarray quality that quantifies the ratio of the intensity distribution of a microarray relative to all other microarrays to be analyzed (McCall et al., 2011). The first method is the relative log expression (RLE), which is calculated by subtracting the median probe set expression estimate across arrays, from each probe set expression estimate in each array (Bolstad et al., 2004). Since most genes are not changed across the arrays, it is expected that these ratios should be around 0 on a log scale (Fig. 3).

The second method is the normalized unscaled standard error (NUSE) that provides a measure of the precision of its expression estimate through the calculation of the ratio of the unscaled standard error of a probe set in a given array, relative to the median standard error of all arrays. Similar to RLE, assuming that most probe sets are not changed across the arrays, it is expected that NUSE ratios are around 1 on a linear scale (Fig. 3). RLE and NUSE values can be displayed in boxplots and summarized with the median and inter-quartile range (IQR). Because both methods rely on multiarray normalized data, values from different batches are not directly comparable (McCall et al., 2011). Recently, a modification of the NUSE metric, called Global NUSE (GNUSE), has been introduced. GNUSE qualifies an individual microarray relative to a balanced sample of all publicly available microarray data on a given platform (McCall et al., 2011). Similar to RLE and NUSE, MA plots identify intensity-dependent biases through pairwise comparison of log-intensity of each array to a reference array, usually corresponding to the average array of all microarrays in the batch (Yang et al., 2002).

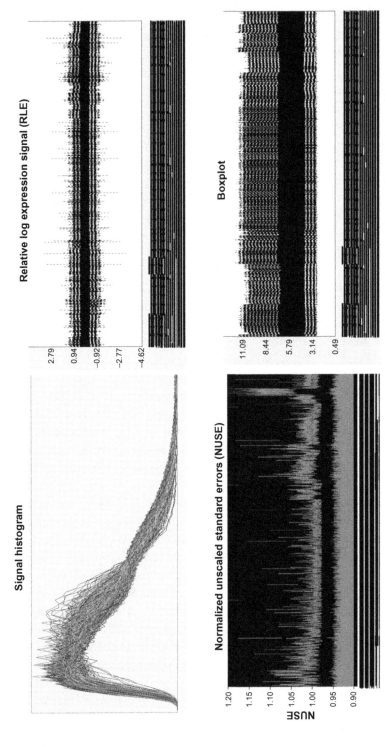

Fig. 3 Methods for assessing homogeneity and detecting outliers in integrated data. These plots were created using over 3000 integrated microarray samples data obtained using Affymetrix HG_U133Plus2 chips.

Several R packages that can perform multiple quality control metrics are also available on the Bioconductor website, of importance are *Simpleaffy*, *AffyPLM,* and *arrayQuality-Metrics* packages (Table 1). *Simpleaffy* performs single-array quality metrics, while *affyPLM* performs multiarray metrics. On the other hand, *arrayQualityMetrics* joins both single-array and multiarray metrics. In addition to Affymetrix data quality control, *arrayQualityMetrics* also performs quality control on Illumina and two-color arrays. An interesting feature in this package is that it provides a detailed report with detected outliers identified in each method. Similarly, *lumi* and *beadarray* are specialized packages for quality assessment of Illumina expression microarrays. Since multiarray metrics is performed after data normalization, it is highly recommended to check RNA quality and single-array data before multiarray quality control so that the latter would not be affected by the low-quality samples.

2.3 Data Preprocessing

After identifying outlier arrays, and after ensuring that the retained samples are all homogeneous and of good quality, raw data from different studies can be joined and analyzed simultaneously. Since platforms differ in their production methods, probes, labeling methods, and hybridization standards, it is necessary that integration be done using data from the same platform (Barnes et al., 2005). Although different methods, including R packages, have been proposed to integrate data from different platforms, it is highly recommended not to go through this step, rather we advise to do the integration on results obtained from each platform after data analysis.

2.3.1 Background Correction

The first step in data preprocessing is to perform background correction to identify non-specific binding of the fluorophore that can be the result of noncomplementary binding or nonbiological sources (noise) (Sánchez and Villa, 2008). In fact, microarray providers integrated different platforms with different methods for background correction based on their production methods (Xie et al., 2009). We provide here a summary for each method but further details can be obtained from the following reference (Gentleman et al., 2005). Affymetrix gene chips use the perfect patch (PM)-mismatch (MM) probe pairs to detect background noise, where PM probes are designed to match exactly the sequence of interest, while the MM probes are designed to contain a single base mismatch at the center base position of the 25-mer oligonucleotide probe. Originally, Affymetrix used Microarray Suite 5 (MAS5) method (Affymetrix Inc, 2004), where the chip is broken into a grid of 16 rectangular regions. For each region the lowest 2% of probe intensities are used to compute a background value that is then used to adjust probe intensities, both MM and PM, based upon a Tukey biweight average of the backgrounds for each of the regions. After adjusting for background noise, the PM intensity is further adjusted for by the MM probe. In theory, MM probes should have lower intensities than the PM probes. However, Irizarry et al. (2003a) showed that around 30% of MM probes have higher intensities than their corresponding PM probes. In consequence, the authors proposed a procedure to improve background correction that uses only the PM intensities using a global model for the distribution of probe intensities derived from empirically motivated statistical models (Irizarry et al., 2003b). This method is now referred to as RMA. The RMA model performs a background correction by fitting a two-component model to the PM intensities that sums up the PM intensity to noise signal. Later, Naef and Magnasco (2003) proposed a new method, known as GCRMA that uses probe sequence information to estimate probe affinity to nonspecific binding by estimating parameters of the position-specific base contributions to the probe affinity. To preserve the use of raw MM intensities, Affymetrix introduced the Ideal Mismatch (IM) model that uses MM intensity for adjustment, only when it is smaller than the PM intensity. If the MM value is larger than the PM value, an idealized value can be estimated based on either the average ratio between PM and MM, or a value slightly smaller than PM (Affymetrix, Inc., 2002). Finally, Li and Wong (2001) proposed a model-based statistical adjustment method that detects outliers and accounts for probe-specific effects in the computation of expression indexes. This method uses both PM and MM in constructing the model and showed reproducible results across different microarray experiments. Illumina and Agilent, on the other hand, perform direct adjustment of probe intensities by subtracting background from foreground measurements to give the final intensity (Ritchie et al., 2011; Zahurak et al., 2007).

2.3.2 Normalization

After background correction, normalization is performed to remove technical artifact differences so that the distribution of intensities is comparable between the arrays. In general, normalization methods are based on the assumption that only a small number of genes are differentially expressed between arrays, and thus, the distribution of probe intensities should be similar. Affymetrix uses the scaling method that is implemented in the MAS5 procedure, where all arrays are scaled to a baseline array corresponding to that having the median of all median intensities (Affymetrix Inc, 2004). Several nonlinear methods have been proposed for data normalization, of importance are cyclic Loess and Quantile normalization. Loess is a method of local regression that is based on MA plot, which normalizes the data by subtracting the difference between the arrays from the average mean difference across arrays (Cleveland and Devlin, 1988; Dudoit et al., 2002; Bolstad et al., 2003). In cyclic Loess, normalizations are carried out in a pairwise manner between each pair of arrays with a repeated process. Quantile normalization, on the other hand, transforms the data by taking the mean quantile across arrays and substituting it as the value of the data item in the original dataset (Bolstad et al., 2003).

2.3.3 Summarization

Affymetrix GeneChip uses a set of probes to measure expression levels of a gene, or an exon in exon arrays. On the other hand, Illumina arrays use technical replicates of one probe for each gene. Summarization is the final stage in preprocessing and is the procedure of combining multiple probe intensities for each probeset into one expression value. Summarization methods are usually implemented in other preprocessing procedures. For instance, MAS5 uses Tukey's biweight method to calculate a robust mean of background corrected log transformed intensity as an expression measure (Wu, 2009). A drawback in this method is that it underestimates the highly consistent probes across arrays. On the other hand, RMA and GCRMA use a linear model that relies on probe efficiency and gene expression on each array to predict the probe and array effects using preprocessed log PM intensity (Wu, 2009). The method then uses median polish to remove the significance of the various elements in the factor model.

2.3.4 Comparison Between Different Procedures

Several studies have been performed to compare between the different methods proposed at different levels. In general, RMA and GCRMA outperform MAS5 and Probe Logarithmic Intensity Error (PLIER). By analyzing microarrays performed on serial dilutions of the same sample, it was shown that preprocessing algorithms that ignore mismatch data, such as RMA and GCRMA, retain the relationship between observed expression and nominal concentrations. In addition, MAS5.0 and PLIER showed the worst precision and highest bias among the different assessed methods (McCall and Almudevar, 2012; Cope et al., 2004). Compared to other methods, GCRMA showed a decrease in the bias in expression measurements without much increase of variance (Wu et al., 2004). However, Giorgi et al. showed that RMA and GCRMA may consistently overestimate sample similarity upon normalization due to polished summarization. Therefore, tRMA is proposed as alternative to RMA/GCRMA that massively reduces interarray correlation artifacts without affecting the detection of differentially expressed genes (Giorgi et al., 2010). Using two different datasets, Bolstad et al. (2003) showed that quantile normalization outperformed all other normalization methods in terms of variance and bias. Finally, it was shown that the preprocessing methods perform differently based on the desired analysis to be done. For instance, it was shown that PM only-based methods outperformed in differential gene expression analysis, while Li–Wong method performed best in coexpression analysis (Harr and Schlötterer, 2006). On the other hand, MAS5 showed better consistency in enrichment analysis (Lim et al., 2007).

The different preprocessing procedures are implemented in the *affy* package (Gautier et al., 2004) for Affymetrix microarray analysis, whereas the *beadarray* package (Ritchie et al., 2011) is available for Illumina BeadChip arrays. Some default functions in *affy* package could be directly implemented for data preprocessing, such as *rma* and *gcrma* functions, while others allow for combination of methods, such as *expresso* and *threestep* functions.

2.4 Batch Effect: A Special Concern During Data Integration

An important issue to resolve during data integration is the adjustment for batch effect before data analysis. Batch effects are nonbiological experimental variations that result when preparing microarray gene expression data from multiple cohorts, array platforms, or different lots (batches) at a time. These variations may lead to nonbiological variations in gene expression that can cause increased false discovery. Therefore, batch effect adjustment is a critical step in data integration to obtain reliable results with low false discovery rate (FDR). Batch effect can be readily visualized using boxplots, unsupervised clustering, and principal component analysis (PCA). A new approach has been also proposed that uses guided PCA (gPCA) to particularly check for batch effect in high dimensional data (Reese et al., 2013).

Several parametric and nonparametric methods have been developed for batch effect adjustment in transcriptomic data such as Support Vector Machines (SVM) (Benito et al., 2004), mean adjustment using threshold scales (PAMR) (Tibshirani et al., 2002), Surrogate variable analysis (SVA) (Leek and Storey, 2007), empirical Bayes methods (ComBat) (Johnson et al., 2007), and finally, using linear models implemented in the *limma* package (Ritchie et al., 2015). Comparisons between methods showed superior performance of certain methods in batch effect adjustment in high dimensional transcriptomic data. Indeed, several studies showed that empirical Bayes algorithm implemented in ComBat outperformed other methods in reducing batch effects while increasing precision and accuracy when applied on both simulated and real data (Johnson et al., 2007; Chen et al., 2011; Kupfer et al., 2012; Lazar et al., 2013; Larsen et al., 2014; Müller et al., 2016). This was also true for using ComBat in small size batches, which raises the reliability of the method in data integration. On the other hand, SVA was recommended for RNA-seq studies (Liu and Markatou, 2016). Both ComBat and SVA are implemented in the Bioconductor package *sva* (Leek and Storey, 2007). However, one should be careful when applying any of the methods on a dataset where the interesting biological factor is confounded with batch effect, which can deflate or fail to remove the batch differences (Soneson et al., 2014; Nygaard et al., 2016) (Fig. 4). In most cases, it is recommended that batch adjustment be done mainly for global visualization of data, while using it with great caution when performing differential gene expression analysis (Nygaard et al., 2016). Alternatively, one may treat the results as an ordered list and select candidates that are the most likely to be true positives (Nygaard et al., 2016). In any case, it is always recommended to ensure a balanced design in which the study groups are evenly distributed across batches (Nygaard et al., 2016).

3. CROSSPLATFORM INTEGRATION

The procedure described in this chapter is mainly adapted to single-platform data integration. However, it might be desirable to perform integration of data that were obtained

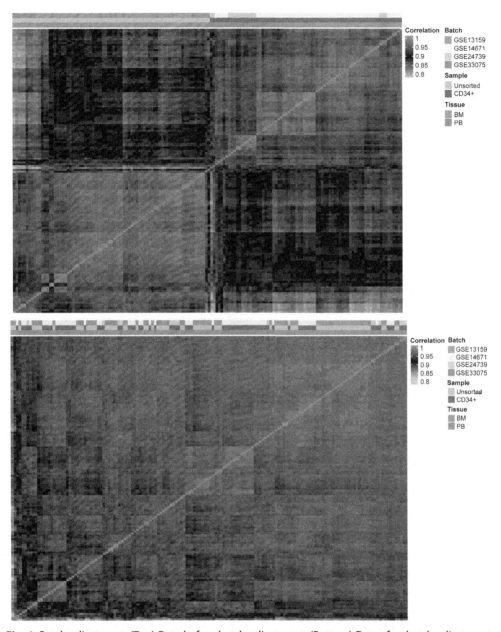

Fig. 4 Batch adjustment. (Top) Data before batch adjustment. (Bottom) Data after batch adjustment. Heat maps were created using 180 integrated microarray data obtained using Affymetrix HG_U133Plus2 chips. Data correspond to 4 microarrays in which batch was highly unbalanced with interesting covariates. As can be seen from the data before batch adjustment (top), correlation between samples is mainly affected by the origin of the samples (unsorted or CD34+), rather than by batch effect. This may lead to indirectly removing biological variations by batch adjustment as can be seen in bottom graph.

using different platforms, e.g., Affymetrix and Illumina. Crossplatform integration (CPI) has been used in coexpression analysis (Nehme et al., 2015; Nehme et al., 2016), functional annotation, and network reconstruction (Zhou et al., 2005; Sîrbu et al., 2010) and was shown to improve gene expression based classification of phenotypes in cancer (Warnat et al., 2005). However, one should be highly cautious in performing such type of data integration as the data from different batches include data with different sample preparation, preprocessing, and gene identification (probe sequences). For such analysis, crossplatform normalization (XPN) is the most recommended method to be used for data normalization and integration (Sîrbu et al., 2010; Shabalin et al., 2008; Rudy and Valafar, 2011).

This method uses local normalization procedure within blocks of similar genes and samples across the platforms (Shabalin et al., 2008). Another approach uses coinertia analysis to identify consensus and divergence between the gene expression profiles from different microarray platforms, and thus, define the genes causing the main trends in the analysis (Culhane et al., 2003). On the other hand, we have recently used a ranking method to identify coexpression patterns in a large group of datasets obtained using different platforms. However, this approach was only used to analyze a specific set of genes and its reliability in global analysis requires further analysis. Finally, WebArrayDB (Xia et al., 2009) and InCroMAP (Wrzodek et al., 2013) are two free platforms that include data repositories for CPI. In addition, several other softwares, besides R/Bioconductor packages, are available to perform CPI, which are reviewed in the following reference (Walsh et al., 2015).

4. CONCLUSION

Transcriptomics has greatly contributed to the systems biology field by transforming molecular biology from the one gene-one function hypothesis approach into large-scale data analysis for inference of networks that generate cellular functions and phenotypic states. In addition, it highly improved biomarker discovery through the simultaneous measure of large numbers of molecular components between subjects. Of importance, a large number of datasets are being added to public repositories each year, which offers the ability for data reanalysis and integration. Data integration is increasingly becoming an essential tool to cope with the growing amounts of data that increase statistical power and reliability of analysis. Despite the profit in using data integration, the approach raises several technical and statistical hurdles that are being addressed by statisticians and computational biologists.

The greatest challenges in horizontal data integration are quality control (Warren and Liu, 2004), data preprocessing (Irizarry et al., 2003a), and batch effect adjustment (Nygaard et al., 2016), which have been addressed by several methods with different levels of precision and bias. In fact, there is not one method that systematically performs best in all cases. Therefore, a combination of different approaches has to be performed at different levels of data preparation in order to obtain the desired data quality for achieving reliable results.

ACKNOWLEDGMENT

This work was supported by the "Fondation de France" and Lebanese National Center of Scientific Research (L-NCSR).

REFERENCES

Affymetrix Inc, 2004. GeneChip Expression Analysis: Data Analysis Fundamentals, Technical Manual. https://www.abdn.ac.uk/ims/documents/data_analysis_fundamentals_manual.pdf.

Affymetrix, 2002. Statistical Algorithms Description Document. Affymetrix, Inc. This is another Manual/Book Available at: http://media.affymetrix.com/support/technical/whitepapers/sadd_whitepaper.pdf.

Barnes, M., Freudenberg, J., Thompson, S., Aronow, B., Pavlidis, P., 2005. Experimental comparison and cross-validation of the Affymetrix and Illumina gene expression analysis platforms. Nucleic Acids Res. 33, 5914–5923.

Barrett, T., Wilhite, S.E., Ledoux, P., Evangelista, C., Kim, I.F., Tomashevsky, M., Marshall, K.A., Phillippy, K.H., Sherman, P.M., Holko, M., Yefanov, A., Lee, H., Zhang, N., Robertson, C.L., Serova, N., Davis, S., Soboleva, A., 2013. NCBI GEO: archive for functional genomics data sets—update. Nucleic Acids Res. 41, D991–D995.

Benito, M., Parker, J., Du, Q., Wu, J., Xiang, D., Perou, C.M., Marron, J.S., 2004. Adjustment of systematic microarray data biases. Bioinforma. Oxf. Engl. 20, 105–114.

Bolstad, B.M., Irizarry, R.A., Astrand, M., Speed, T.P., 2003. A comparison of normalization methods for high density oligonucleotide array data based on variance and bias. Bioinforma. Oxf. Engl. 19, 185–193.

Bolstad, B.M., Collin, F., Simpson, K.M., Irizarry, R.A., Speed, T.P., 2004. Experimental design and low-level analysis of microarray data. Int. Rev. Neurobiol. 60, 25–58.

Chen, C., Grennan, K., Badner, J., Zhang, D., Gershon, E., Jin, L., Liu, C., 2011. Removing batch effects in analysis of expression microarray data: an evaluation of six batch adjustment methods. PLoS One 6, e17238.

Cleveland, W.S., Devlin, S.J., 1988. Locally weighted regression: an approach to regression analysis by local fitting. J. Am. Stat. Assoc. 83, 596–610.

Colaprico, A., Silva, T.C., Olsen, C., Garofano, L., Cava, C., Garolini, D., Sabedot, T.S., Malta, T.M., Pagnotta, S.M., Castiglioni, I., Ceccarelli, M., Bontempi, G., Noushmehr, H., 2016. TCGAbiolinks: an R/Bioconductor package for integrative analysis of TCGA data. Nucleic Acids Res. 44, e71.

Cope, L.M., Irizarry, R.A., Jaffee, H.A., Wu, Z., Speed, T.P., 2004. A benchmark for Affymetrix GeneChip expression measures. Bioinforma. Oxf. Engl. 20, 323–331.

Culhane, A.C., Perrière, G., Higgins, D.G., 2003. Cross-platform comparison and visualisation of gene expression data using co-inertia analysis. BMC Bioinforma. 4, 59.

Davis, S., Meltzer, P.S., 2007. GEOquery: a bridge between the gene expression omnibus (GEO) and BioConductor. Bioinforma. Oxf. Engl. 23, 1846–1847.

Dudoit, S., Yang, Y.H., Callow, M.J., Speed, T.P., 2002. Statistical methods for identifying differentially expressed genes in replicated cDNA microarray experiments. Stat. Sin. 12, 111–139.

Gautier, L., Cope, L., Bolstad, B.M., Irizarry, R.A., 2004. Affy—analysis of Affymetrix GeneChip data at the probe level. Bioinforma. Oxf. Engl. 20, 307–315.

Gentleman, R., Carey, V., Huber, W., Irizarry, R., Dudoit, S. (Eds.), 2005. Bioinformatics and Computational Biology Solutions Using R and Bioconductor. Springer. https://doi.org/10.1007/0-387-29362-0. http://www.springer.com/us/book/9780387251462.

Giorgi, F.M., Bolger, A.M., Lohse, M., Usadel, B., 2010. Algorithm-driven artifacts in median polish summarization of microarray data. BMC Bioinforma. 11, 553.

Hamid, J.S., Hu, P., Roslin, N.M., Ling, V., Greenwood, C.M.T., Beyene, J., 2009. Data integration in genetics and genomics: methods and challenges. Hum. Genomics Proteomics. https://doi.org/10.4061/2009/869093.

Harr, B., Schlötterer, C., 2006. Comparison of algorithms for the analysis of Affymetrix microarray data as evaluated by co-expression of genes in known operons. Nucleic Acids Res. 34, e8.

Huber, W., Carey, V.J., Gentleman, R., Anders, S., Carlson, M., Carvalho, B.S., Bravo, H.C., Davis, S., Gatto, L., Girke, T., Gottardo, R., Hahne, F., Hansen, K.D., Irizarry, R.A., Lawrence, M., Love, M.I., MacDonald, J., Obenchain, V., Oleś, A.K., Pagès, H., Reyes, A., Shannon, P., Smyth, G.K., Tenenbaum, D., Waldron, L., Morgan, M., 2015. Orchestrating high-throughput genomic analysis with Bioconductor. Nat. Methods 12, 115–121.

Irizarry, R.A., Hobbs, B., Collin, F., Beazer-Barclay, Y.D., Antonellis, K.J., Scherf, U., Speed, T.P., 2003a. Exploration, normalization, and summaries of high density oligonucleotide array probe level data. Biostat. Oxf. Engl. 4, 249–264.

Irizarry, R.A., Bolstad, B.M., Collin, F., Cope, L.M., Hobbs, B., Speed, T.P., 2003b. Summaries of Affymetrix GeneChip probe level data. Nucleic Acids Res. 31, e15.

Johnson, W.E., Li, C., Rabinovic, A., 2007. Adjusting batch effects in microarray expression data using empirical Bayes methods. Biostat. Oxf. Engl. 8, 118–127.

Kauffmann, A., Rayner, T.F., Parkinson, H., Kapushesky, M., Lukk, M., Brazma, A., Huber, W., 2009a. Importing ArrayExpress datasets into R/Bioconductor. Bioinforma. Oxf. Engl. 25, 2092–2094.

Kauffmann, A., Gentleman, R., Huber, W., 2009b. arrayQualityMetrics—a bioconductor package for quality assessment of microarray data. Bioinforma. Oxf. Engl. 25, 415–416.

Kupfer, P., Guthke, R., Pohlers, D., Huber, R., Koczan, D., Kinne, R.W., 2012. Batch correction of microarray data substantially improves the identification of genes differentially expressed in rheumatoid arthritis and osteoarthritis. BMC Med. Genet. 5, 23.

Larsen, M.J., Thomassen, M., Tan, Q., Sørensen, K.P., Kruse, T.A., 2014. Microarray-based RNA profiling of breast cancer: batch effect removal improves cross-platform consistency. BioMed Res. Int. https://doi.org/10.1155/2014/651751.

Lazar, C., Meganck, S., Taminau, J., Steenhoff, D., Coletta, A., Molter, C., Weiss-Solís, D.Y., Duque, R., Bersini, H., Nowé, A., 2013. Batch effect removal methods for microarray gene expression data integration: a survey. Brief. Bioinform. 14, 469–490.

Leek, J.T., Storey, J.D., 2007. Capturing heterogeneity in gene expression studies by surrogate variable analysis. PLoS Genet. 3, 1724–1735.

Li, C., Wong, W.H., 2001. Model-based analysis of oligonucleotide arrays: expression index computation and outlier detection. Proc. Natl. Acad. Sci. USA 98, 31–36.

Lim, W.K., Wang, K., Lefebvre, C., Califano, A., 2007. Comparative analysis of microarray normalization procedures: effects on reverse engineering gene networks. Bioinformatics 23, i282–i288.

Liu, Q., Markatou, M., 2016. Evaluation of methods in removing batch effects on RNA-seq data. Infect. Dis. Transl. Med. 2, 3–9.

McCall, M.N., Almudevar, A., 2012. Affymetrix GeneChip microarray preprocessing for multivariate analyses. Brief. Bioinform. 13, 536–546.

McCall, M.N., Murakami, P.N., Lukk, M., Huber, W., Irizarry, R.A., 2011. Assessing affymetrix GeneChip microarray quality. BMC Bioinforma. 12, 137.

Müller, C., Schillert, A., Röthemeier, C., Trégouët, D.-A., Proust, C., Binder, H., Pfeiffer, N., Beutel, M., Lackner, K.J., Schnabel, R.B., Tiret, L., Wild, P.S., Blankenberg, S., Zeller, T., Ziegler, A., 2016. Removing batch effects from longitudinal gene expression-quantile normalization plus ComBat as best approach for microarray transcriptome data. PLoS One 11, e0156594.

Naef, F., Magnasco, M.O., 2003. Solving the riddle of the bright mismatches: labeling and effective binding in oligonucleotide arrays. Phys. Rev. E Stat. Nonlin. Soft Matter Phys. 68, 011906.

Nehme, A., Cerutti, C., Dhaouadi, N., Gustin, M.P., Courand, P.-Y., Zibara, K., Bricca, G., 2015. Atlas of tissue renin-angiotensin-aldosterone system in human: a transcriptomic meta-analysis. Sci. Rep. 5, 10035.

Nehme, A., Cerutti, C., Zibara, K., 2016. Transcriptomic analysis reveals novel transcription factors associated with renin-angiotensin-aldosterone system in human atheroma. Hypertension. https://doi.org/10.1161/HYPERTENSIONAHA.116.08070.

Nygaard, V., Rødland, E.A., Hovig, E., 2016. Methods that remove batch effects while retaining group differences may lead to exaggerated confidence in downstream analyses. Biostat. Oxf. Engl. 17, 29–39.

Parkinson, H., Kapushesky, M., Kolesnikov, N., Rustici, G., Shojatalab, M., Abeygunawardena, N., Berube, H., Dylag, M., Emam, I., Farne, A., Holloway, E., Lukk, M., Malone, J., Mani, R., Pilicheva, E., Rayner, T.F., Rezwan, F., Sharma, A., Williams, E., Bradley, X.Z., Adamusiak, T., Brandizi, M., Burdett, T., Coulson, R., Krestyaninova, M., Kurnosov, P., Maguire, E., Neogi, S.G., Rocca-Serra, P., Sansone, S.-A., Sklyar, N., Zhao, M., Sarkans, U., Brazma, A., 2009. ArrayExpress update—from an archive of functional genomics experiments to the atlas of gene expression. Nucleic Acids Res. 37, D868–D872.

Reese, S.E., Archer, K.J., Therneau, T.M., Atkinson, E.J., Vachon, C.M., de Andrade, M., Kocher, J.-P.A., Eckel-Passow, J.E., 2013. A new statistic for identifying batch effects in high-throughput genomic data that uses guided principal component analysis. Bioinformatics 29, 2877–2883.

Ritchie, M.E., Dunning, M.J., Smith, M.L., Shi, W., Lynch, A.G., 2011. BeadArray expression analysis using bioconductor. PLoS Comput. Biol. https://doi.org/10.1371/journal.pcbi.1002276.

Ritchie, M.E., Phipson, B., Wu, D., Hu, Y., Law, C.W., Shi, W., Smyth, G.K., 2015. Limma powers differential expression analyses for RNA-sequencing and microarray studies. Nucleic Acids Res. 43, e47.

Rudy, J., Valafar, F., 2011. Empirical comparison of cross-platform normalization methods for gene expression data. BMC Bioinforma. 12, 467.

Sánchez, A., Villa, M.C.R.D., 2008. A tutorial review of microarray data analysis, pp. 1–55. http://www.ub.edu/stat/docencia/bioinformatica/microarrays/ADM/slides/A_Tutorial_Review_of_Microarray_data_Analysis_17-06-08.pdf.

Shabalin, A.A., Tjelmeland, H., Fan, C., Perou, C.M., Nobel, A.B., 2008. Merging two gene-expression studies via cross-platform normalization. Bioinformatics 24, 1154–1160.

Sîrbu, A., Ruskin, H.J., Crane, M., 2010. Cross-platform microarray data normalisation for regulatory network inference. PLoS One 5, e13822.

Soneson, C., Gerster, S., Delorenzi, M., 2014. Batch effect confounding leads to strong bias in performance estimates obtained by cross-validation. PLoS One. https://doi.org/10.1371/journal.pone.0100335.

Tibshirani, R., Hastie, T., Narasimhan, B., Chu, G., 2002. Diagnosis of multiple cancer types by shrunken centroids of gene expression. Proc. Natl. Acad. Sci. USA 99, 6567–6572.

Tomczak, K., Czerwińska, P., Wiznerowicz, M., 2015. The cancer genome atlas (TCGA): an immeasurable source of knowledge. Contemp. Oncol. Poznan Pol. 19, A68–A77.

Tseng, G.C., Ghosh, D., Feingold, E., 2012. Comprehensive literature review and statistical considerations for microarray meta-analysis. Nucleic Acids Res. 40, 3785–3799.

Vizcaíno, J.A., Csordas, A., del-Toro, N., Dianes, J.A., Griss, J., Lavidas, I., Mayer, G., Perez-Riverol, Y., Reisinger, F., Ternent, T., Xu, Q.-W., Wang, R., Hermjakob, H., 2016. 2016 update of the PRIDE database and its related tools. Nucleic Acids Res. 44, D447–D456.

Walsh, C.J., Hu, P., Batt, J., Dos Santos, C.C., 2015. Microarray meta-analysis and cross-platform normalization: integrative genomics for robust biomarker discovery. Microarrays 4, 389–406.

Warnat, P., Eils, R., Brors, B., 2005. Cross-platform analysis of cancer microarray data improves gene expression based classification of phenotypes. BMC Bioinforma. 6, 265.

Warren, L., Liu, B., 2004. Comparison of Normalization Methods for cDNA Microarrays. In: Johnson, K.F., Lin, S.M. (Eds.), Methods of Microarray Data Analysis III. Springer, Boston, MA. https://doi.org/10.1007/0-306-48354-8_8. (Accessed 21 October 2017).

Wilhelm, M., Schlegl, J., Hahne, H., Gholami, A.M., Lieberenz, M., Savitski, M.M., Ziegler, E., Butzmann, L., Gessulat, S., Marx, H., Mathieson, T., Lemeer, S., Schnatbaum, K., Reimer, U., Wenschuh, H., Mollenhauer, M., Slotta-Huspenina, J., Boese, J.-H., Bantscheff, M., Gerstmair, A., Faerber, F., Kuster, B., 2014. Mass-spectrometry-based draft of the human proteome. Nature 509, 582–587.

Wilson, C.L., Miller, C.J., 2005. Simpleaffy: a BioConductor package for Affymetrix Quality Control and data analysis. Bioinforma. Oxf. Engl. 21, 3683–3685.

Wishart, D.S., Jewison, T., Guo, A.C., Wilson, M., Knox, C., Liu, Y., Djoumbou, Y., Mandal, R., Aziat, F., Dong, E., Bouatra, S., Sinelnikov, I., Arndt, D., Xia, J., Liu, P., Yallou, F., Bjorndahl, T., Perez-Pineiro, R., Eisner, R., Allen, F., Neveu, V., Greiner, R., Scalbert, A., 2013. HMDB 3.0—the human metabolome database in 2013. Nucleic Acids Res. 41, D801–D807.

Wrzodek, C., Eichner, J., Büchel, F., Zell, A., 2013. InCroMAP: Integrated analysis of cross-platform microarray and pathway data. Bioinforma. Oxf. Engl. 29, 506–508.

Wu, Z., 2009. A review of statistical methods for preprocessing oligonucleotide microarrays. Stat. Methods Med. Res. 18, 533–541.

Wu, Z., Irizarry, R.A., Gentleman, R., Martinez-Murillo, F., Spencer, F., 2004. A model-based background adjustment for oligonucleotide expression arrays. J. Am. Stat. Assoc. 99, 909–917.

Xia, X.-Q., McClelland, M., Porwollik, S., Song, W., Cong, X., Wang, Y., 2009. WebArrayDB: cross-platform microarray data analysis and public data repository. Bioinformatics 25, 2425–2429.

Xie, Y., Wang, X., Story, M., 2009. Statistical methods of background correction for Illumina BeadArray data. Bioinformatics 25, 751–757.

Yang, Y.H., Dudoit, S., Luu, P., Lin, D.M., Peng, V., Ngai, J., Speed, T.P., 2002. Normalization for cDNA microarray data: a robust composite method addressing single and multiple slide systematic variation. Nucleic Acids Res. 30, e15.

Zahurak, M., Parmigiani, G., Yu, W., Scharpf, R.B., Berman, D., Schaeffer, E., Shabbeer, S., Cope, L., 2007. Pre-processing Agilent microarray data. BMC Bioinforma. 8, 142.

Zhou, X.J., Kao, M.-C.J., Huang, H., Wong, A., Nunez-Iglesias, J., Primig, M., Aparicio, O.M., Finch, C.E., Morgan, T.E., Wong, W.H., 2005. Functional annotation and network reconstruction through cross-platform integration of microarray data. Nat. Biotechnol. 23, 238–243.

CHAPTER 2

Proteomics and Protein Interaction in Molecular Cell Signaling Pathways

Hassan Pezeshgi Modarres*, Mohammad R.K. Mofrad*,†
*Molecular Cell Biomechanics Laboratory, Departments of Bioengineering and Mechanical Engineering, University of California Berkeley, Berkeley, CA, United States
†Physical Biosciences Division, Lawrence Berkeley National Lab, Berkeley, CA, United States

1. INTRODUCTION

The genome sequence of many organisms can now be found in freely available online databases. Such resources provide valuable data that can lead to improving our knowledge about function of genes and proteins when appropriate analysis tools are employed. Wilkins et al. introduced the term "proteome" initially to describe the complement of protein to the genome (Wilkins et al., 1996). Proteomics has since emerged as an important research field in biology where different properties of proteins are studied, including their composition, expression level in response to different environmental conditions, their posttranslational modifications, function, structure, and interactions that govern the cell's different biological processes, to gain a universal perspective of the cell's machinery at the level of proteins (Samaj and Thelen, 2007; Anderson and Anderson, 1998; Blackstock and Weir, 1999). Comparing with genomics, proteomics gives a better operational picture of an organism's functions. Although the genome of an organism is more or less fixed, the expression profile of proteins varies and strongly depends on the environmental conditions (Chandrasekhar et al., 2014). While the human genome encodes 26,000–31,000 protein sequences, the number of different variants and products of proteins such as posttranslational modifications and protein splices is estimated to be one million (Wilkins et al., 1999; Godovac-Zimmermann and Brown, 2001). Moreover, a huge amount of the encoded data within the genome can also lead us to the proteomic information such as protein localization, trafficking, phosphorylation, and protein-protein interactions (Alaoui-Jamali and Xu, 2006). In this chapter, we will give a brief overview of state-of-the-art proteomics studies with an emphasis on the computational tools and techniques in proteomics studies.

2. EXPERIMENTAL TECHNIQUES

To understand the function of proteins, different high-throughput experimental techniques have been developed that can capture the protein expression levels at different

Leveraging Biomedical and Healthcare Data
https://doi.org/10.1016/B978-0-12-809556-0.00002-2
17

environmental conditions. These include mRNA expression profiling using microarray technologies and protein expression profiling by mass spectrometry and two-dimensional electrophoresis (2DE) (Jollès and Jörnvall, 2013) (see Tables 1 and 2). Proteomics studies often feature several complicated experimental and computational steps including sample preparation, detection and measurement techniques like mass spectroscopy, and data processing and analysis (Jollès and Jörnvall, 2013). The main aim of the sample preparation step is to reduce the complexity by depletion of the most abundant proteins and enrichment of lower abundant proteins. For proteomics studies, different approaches have been utilized including gel-based techniques (one- and two-dimensional) (Van den Bergh and Arckens, 2005), isotope-coded affinity tag (ICAT) (Gygi et al., 1999), multidimensional protein identification technology (Nedelkov and Nelson, 2006), isobaric tagging (iTRAQ) (Ross et al., 2004), stable isotope labeling by/with amino acids in cell culture (SILAC) (Ong et al., 2002), two-dimensional difference gel electrophoresis (2D-

Table 1 Databases for proteomics techniques (gel electrophoresis (GE) and mass spectrometry (MS)

Database name	Description	URL
WORLD-2DPAGE Constellation	GE	http://www.world-2dpage.expasy.org/
Delta2Da	GE	www.decodon.com/Solutions/Delta2D.html
GD Impressionista	GE	www.genedata.com/productsgell/Gellab.html
Investigator HT PC Analyzera	GE	www.genomicsolutions.com/proteomics/ 2dgelanal.html
Phortix 2Da	GE	www.phortix.com/products/2d_products.htm
Z3 2D-Gel Analysis Systema	GE	www.2dgels.com
Global Proteome Machine Database (GPMDB)	MS	http://www.thegpm.org/GPMDB/index.html
PeptideAtlas	MS	http://www.peptideatlas.org/
Peptidome	MS	http://www.ncbi.nlm.nih.gov/peptidome/
Mascot	MS	www.matrixscience.com
MassSearch	MS	www.Cbrg.inf.ethz/Server/MassSearch.html
MS-FIT	MS	www.Prospector.ucsf.edu
PeptIdent	MS	www.expasy.ch/tools/peptident.html

Table 2 Expression databases

Name	URL
ArrayExpress	http://www.ebi.ac.uk/arrayexpress
SMD	http://genome-www5.stanford.edu
CGAP	http://cgap.nci.nih.gov
GEO	http://www.ncbi.nlm.nih.gov/geo

DIGE) (Klose et al., 2002), shotgun proteomics (Wolters et al., 2001), protein microarrays (Melton, 2004), label-free quantification of high mass resolution LC-MS data (Mueller et al., 2007), Western blot assays (Schulz et al., 2007), and multiple reaction monitoring (MRM) assay (Stahl-Zeng et al., 2007).

However, experimental proteomics studies and technologies still deal with serious challenges. For example, because of the high hydrophobicity of integral membrane proteins, during the isoelectric focusing step of 2DE method, protein aggregation happens and limits this method to proteins with low hydrophobicity or membrane-associated proteins while in the genome 20%–30% of genes are encoding integral membrane proteins with numerous functions (Team, 2002; Zuo and Speicher, 2000). To overcome this limitation, different methods like cell-surface biotinylation (Castronovo et al., 2006; Scheurer et al., 2005), chemical-tagging methods (Zhang et al., 2004), and membrane solubilization approaches using organic solvents (Blonder et al., 2002), detergents (Han et al., 2001), and organic acids (Washburn et al., 2001) have been developed.

As another example, biomarker discovery and serum proteomics deal with highly variable states of modification and concentration of some proteins within the human plasma that contains more than 10,000 different proteins while many are shed by cell or secreted at different pathological or physiological conditions (Anderson and Anderson, 2002). By techniques like heparin chromatography coupled with protein G, the dynamic range and complexity of protein concentrations can be overcome by prefractionation and depleting the highly abundant proteins (Lei et al., 2008). IgG can be removed by affinity chromatography (Björck and Kronvall, 1984) and albumin can be removed by isoelectric trapping (Rothemund et al., 2003) immunoaffinity columns (Govorukhina et al., 2003), peptide affinity chromatography (Sato et al., 2002), and dye-ligand chromatography (Travis et al., 1976).

3. COMPUTATIONAL METHODS

With the vast amount of data coming out of proteomics studies, new challenges and opportunities have emerged. Dealing with how to store, search, and retain such big data is a major challenge (Smith et al., 2007). On the other hand, using such data provides unique opportunities for computational scientists for training or establishing in silico predictive models that make it possible to address important questions that proteomics is dealing with, for example what is the function of a newly sequenced protein? Where is it located within the cell? What are the posttranslational modification sites? What is the relationship between protein sequence, structure, and functions? How can we compare protein sequences and structures? How is it possible to form protein families that share a common function and find the family that a sequence with unknown function belongs to? How is it possible to compare a newly sequenced protein against a library of millions of protein sequences? Which proteins are potentially candidates to interact with a target sequence? Currently available technologies are either unable or too costly to address these questions. In the following, we will explore the main challenges that computational proteomics is dealing with.

3.1 Sequences Databases and Analysis

During genomics studies of different species, a number of protein sequences are often introduced that have never been studied in vivo and their functions are not known. One of the main challenges in computational proteomics is assigning functions to such newly discovered protein sequences. Function prediction algorithms are basically following a simple rule: if sequence X is similar to sequence Y and the function of protein Y is known, the function of protein X is probably similar to that of protein Y. Therefore, in addition to databases containing protein sequences with assigned functions, accurate and efficient algorithms and tools that can make comparisons between protein sequences are an essential part of computational proteomics. During the last decades, such methods have been successfully used to characterize protein sequences. Table 3 shows databases for protein sequences and their corresponding functions, if reported. As stated, the first step to predict features of a target protein is comparing it with the currently available sequences in the database with assigned features such as stability, active/binding sites, and posttranslational modification sites.

However, a full comparison between the target sequence and all of the available sequences within databases is computationally costly, necessitating faster searching

Table 3 Protein sequence databases

Database name	URL
Reference Sequence (RefSeq)	http://www.ncbi.nlm.nih.gov/RefSeq/
Entrez Protein	http://www.ncbi.nlm.nih.gov/sites/
UniProt Knowledgebase (UniProtKB)	http://www.uniprot.org/help/uniprotkb
UniProt Archive (UniParc)	http://www.uniprot.org/help/uniparc
UniProt Reference Clusters (UniRef)	http://www.uniprot.org/help/uniref
EMBL	http://www.ebi.ac.uk/embl
GenBank	http://www.ncbi.nlm.nih.gov/Genbank
DDBJ	http://www.ddbj.nig.ac.jp
Genome Reviews	http://www.ebi.ac.uk/GenomeReviews
IMGT Databases	http://www.ebi.ac.uk/imgt
IPD-MHC	http://www.ebi.ac.uk/ipd/mhc
GPCRDB	http://www.gpcr.org
RDP	http://rdp.cme.msu.edu
TRANSFAC	http://www.gene-regulation.com
EPD	http://www.epd.isb-sib.ch
HIVdb	http://hivdb.stanford.edu
REBASE	http://rebase.neb.com

approaches. BLAST (Basic Local Alignment Search Tool) (Altschul et al., 1990) search algorithm is implemented to search for a target sequence over a database to find similar sequences within the database and assign a score to each sequence comparison. This method was published in 1990 and has been cited more than 70,100 times (according to Google Scholar). However, BLAST is not the only available sequence database search. BLAT (Blast Like Alignment Tool) (Kent, 2002) is considerably faster than BLAST but is not as flexible as BLAST while BLAST can find much more remote matches but BLAT needs exact or near-exact matches. Hidden Markov Model matching is yet another alternative algorithm that is used in protein family database (Pfam) as implemented in tools such as HMMER (Finn et al., 2015) (see Table 4). Tables 4 presents a list of sequence search and alignment methods.

Sequence alignment involves arranging the protein (or DNA or RNA) sequences to obtain the highest matching residues among sequences. The detected similarity between the sequences, if any, can be a consequence of evolutionary, functional, or structural relationships. In general, two types of alignments are conducted, pairwise and multiple alignments, for two and multiple sequences, respectively (see Tables 5 and 6). For two sequences with a common ancestor, mismatches can represent point mutation and gaps can represent insertion of deletion mutations. Highly conserved residues in the alignment, in particular regions of protein sequences, may have structural or functional ramifications in the protein.

The sequence alignment could feature either local or global alignment strategies. The global alignment approaches are better suited when the sequences are generally similar and of comparable sizes. For example, Needleman-Wunsch is one of the most used algorithms for global alignment based on dynamic programming. On the other hand, when dealing with largely dissimilar sequences containing similar motifs or regions, local alignment techniques are more useful. Smith-Waterman algorithm, also based on dynamic programming, is a local alignment method (Baxevanis and Ouellette, 2004).

3.2 Protein Feature Prediction Using Protein Sequence

Similarities among protein sequences are reminiscent of homology and convergent evolution via common ancestry and/or selective pressure, respectively. Hence, finding similar patterns within protein sequences can help us to annotate functional and structural properties for a given protein sequence. Using sequence alignments makes it possible to detect similarities raised by common ancestry while machine-learning methods such as support vector machines (SVMs) and artificial neural networks are used to detect similarities caused by common selective pressure (Bujnicki, 2008; Juncker et al., 2009). Annotation of different functions of a protein using its sequence in comparison to sequences of known functions is generally performed by recognition of global patterns, such as amino acid pair frequencies or composition, or local patterns, such as conserved

Table 4 Sequence database search tools

Name	Description	URL
BLAST	Local search with fast k-tuple heuristic (Basic Local Alignment Search Tool)	https://blast.ncbi.nlm.nih.gov/Blast.cgi
PSI-BLAST	Position-specific iterative BLAST	https://blast.ncbi.nlm.nih.gov/Blast.cgi?PAGE=Proteins&PROGRAM=blastp&BLAST_PROGRAMS=blastp&PAGE_TYPE=BlastSearch&SHOW_DEFAULTS=on&LINK_LOC=blasthome
ScalaBLAST	Highly parallel Scalable BLAST	https://omics.pnl.gov/software/ScalaBLAST.php
CS-BLAST	More sensitive than BLAST, FASTA, and SSEARCH	http://toolkit.lmb.uni-muenchen.de/cs_blast
DIAMOND	Up to 20,000 times the speed of BLAST, with high sensitivity	http://ab.inf.uni-tuebingen.de/software/diamond/
CUDASW++	For Smith-Waterman protein database searches that takes advantage of the massively parallel CUDA architecture of NVIDIA GPUs to perform sequence searches 10–50× faster than NCBI BLAST	http://cudasw.sourceforge.net/homepage.htm#latest
HMMER	Biosequence analysis using profile hidden Markov models	http://hmmer.org/
FASTA	Local search with fast k-tuple heuristic, faster but less sensitive than BLAST	https://www.ebi.ac.uk/Tools/sss/fasta/
IDF	Inverse Document Frequency Weighted Genomic Sequence Retrieval	http://www.cs.uni.edu/~okane/source/IDF/idf.html
SWIMM	Smith-Waterman implementation for Intel Multicore and Manycore architectures	https://github.com/enzorucci/SWIMM
SWAPHI	Parallelized and accelerated Smith-Waterman protein database search	http://www.ebi.ac.uk/Tools/services/web/toolform.ebi?tool=fasta&program=ssearch&context=protein
SAM	Global local search with profile Hidden Markov models, more sensitive than PSI-BLAST	https://web.archive.org/web/20080509161215/http://www.cse.ucsc.edu/research/compbio/sam.html

Table 5 Pairwise alignment tools

Name	Description	URL
ACANA	Fast and heuristic	http://bioinformatics.joyhz.com/ ACANA/
AlignMe	For membrane proteins	http://www.bioinfo.mpg.de/AlignMe
Bioconductor	Dynamic programming	http://www.bioconductor.org/
DOTLET	Dot-plot	http://www.isrec.isb-sib.ch/java/dotlet/ Dotlet.html
NW-align	Standard Needleman-Wunsch dynamic programming	https://zhanglab.ccmb.med.umich.edu/ NW-align/
Matcher	Waterman-Eggert local alignment	https://galaxy.pasteur.fr/?form=matcher
Needle	Needleman-Wunsch dynamic programming	https://www.ebi.ac.uk/Tools/psa/ emboss_needle/
ProbA	Stochastic partition function sampling	http://www.tbi.univie.ac.at/~ulim/ probA/index.html
SABERTOOTH	Uses predicted connectivity profiles	http://www.fkp.tu-darmstadt.de/ sabertooth_source/
SSEARCH	Local (Smith-Waterman) alignment with statistics	http://www.ebi.ac.uk/Tools/services/ web/toolform.ebi?tool=fasta& program=ssearch&context=protein

regions and motifs, a combination of these examinations (Juncker et al., 2009). The predictions can be in three levels, namely, the whole sequence level, sequence segment level (motifs), and individual residue level (e.g., binding site residues). In general, functional annotations, based on the whole sequence similarity, are accurate only when the sequences are highly similar (Devos and Valencia, 2000). However, for the cases that the pair of proteins does not show an overall similarity, the conserved functional motifs within the alignment can be used for function annotation of target proteins. Therefore, methods such as hidden Markov models, PSI-BLAST (Altschul et al., 1997), ConFunc (Wass and Sternberg, 2008), and PFP (Hawkins et al., 2006) are developed for detection of such functional motifs. At the residue level, functional information such as ligand binding and catalytic binding residues can be detected. Catalytic Site Atlas (Porter et al., 2004) and Firestar (López et al., 2007) are useful tools for design of such predictors. Different tools have been developed to predict phosphorylation sites on proteins such asNetPhosK (Blom et al., 2004), Scansite (Obenauer et al., 2003), NetworKIN (Linding et al., 2007), and NetPhorest (Miller et al., 2008) and for Glycosylation prediction such as NetOGlyc (Julenius et al., 2005), NetCGlyc (Julenius, 2007). Tools also have been designed to predict subcellular localization including WoLF PSORT (Horton et al., 2007), BaCelLo (Pierleoni et al., 2006), TargetP (Emanuelsson et al., 2007), and LOCTree (Nair and Rost, 2002). Particularly, the prediction of the transmembrane domain of

Table 6 Multiple sequence alignment tools

Name	Description	URL
ClustalW	Progressive alignment	https://embnet.vital-it.ch/software/ClustalW.html
MUSCLE	Progressive-iterative alignment	http://www.drive5.com/muscle/
T-Coffee	More sensitive progressive alignment	http://www.tcoffee.org/
Compass	Comparison of multiple protein sequence alignments with assessment of statistical significance	http://prodata.swmed.edu/compass
DECIPHER	Progressive-iterative alignment	http://decipher.cee.wisc.edu/Alignment.html
FAMSA	Progressive alignment for large number of protein sequences	http://sun.aei.polsl.pl/REFRESH/index.php?page=projects&project=famsa&subpage=about
MSA	Dynamic programming	https://www.ncbi.nlm.nih.gov/CBBresearch/Schaffer/msa.html
MULTALIN	Dynamic programming-clustering	http://multalin.toulouse.inra.fr/multalin/
PicXAA	Nonprogressive alignment	http://gsp.tamu.edu/picxaa/index.html
Probalign	Probabilistic/consistency with partition function probabilities	http://probalign.njit.edu/probalign/login
PSAlign	Alignment preserving nonheuristic	http://faculty.cs.tamu.edu/shsze/psalign/
SAGA	Uses genetic algorithm	http://www.tcoffee.org/Projects/saga/
GLProbs	Adaptive pair-Hidden Markov Model based approach	https://sourceforge.net/projects/glprobs/?source=navbar

transmembrane proteins via protein sequence analysis has been one of the old fields of interest using tools like Phobius (Melen et al., 2003), HMMTOP (Viklund and Elofsson, 2004), PONGO (Amico et al., 2006).

3.3 Structure Databases and Analysis

Homologous proteins usually share similar structures, even if their sequences are not highly similar. Comparing protein structures can help us to find similar proteins that potentially show similar functions. In addition, protein structure analysis is a useful way to study proteins' interacting partners (proteins, DNA, RNA, and small-molecule ligands) as well as the mechanism of their interactions (Xu et al., 2007). Table 7 shows a list of databases containing protein 3D structures that are provided by experimental techniques such as X-ray crystallography and NMR.

Table 7 Protein structure databases

Database name	URL
Worldwide Protein Data Bank (wwPDB)	http://www.wwpdb.org/
Molecular Modeling Database (MMDB)	http://www.ncbi.nlm.nih.gov/sites/entrez?db=structure
ModBase	http://www.modbase.compbio.ucsf.edu/modbase-cgi/index.cgi
SWISS-MODEL Repository	http://www.swissmodel.expasy.org/repository/
Protein Folding Database (PFD)	http://www.pfd.med.monash.edu.au/
KineticDB	http://www.kineticdb.protres.ru/db/index.pl
CATH	http://www.cathdb.info/
Structural Classification Of Proteins (SCOP)	http://www.scop.mrc-lmb.cam.ac.uk/scop/
SUPERFAMILY	http://www.supfam.org/SUPERFAMILY/
RESID	http://www.ebi.ac.uk/RESID/
Phospho3D	http://www.cbm.bio.uniroma2.it/phospho3d/

When no experimentally derived 3D structure is available for a given protein, resources featuring databases of theoretical models and protein 3D modeling tools are useful. For example, 3DCrunch (Guex and Peitsch, 1997) and MODBASE (Pieper et al., 2002) provide theoretical models generated using databases such as TrEMBL and SWISS-PROT records (see Table 6). When the protein of interest is not available on these databases, then tools such as SWISS-MODEL (Guex and Peitsch, 1997) or MODELLER (Kihara, 2016) can be used to perform homology modeling and generate a 3D structure by fitting sequence of the target protein on a template. In addition to homology modeling, fold recognition and ab initio modeling are two commonly used approaches to protein structure modeling (see Table 8). An important advantage of 3D modeling of proteins is that we can find structurally and functionally important amino acids and thereby refine the sequence alignment focusing on the important residues.

Like protein sequence comparisons, structure comparisons are helpful in the analysis of protein functions. Efficient protein structure search programs can identify proteins with similar structures to the structure of the target protein without shared evolutionary ancestry in databases. In addition, one can find candidate binding site residues on the protein of interest by comparing it with structure of a similar protein in binding state to another biomolecule such as DNA (Seitz et al., 2014). A list of protein 3D structure comparison tools is presented in Table 9.

3.4 Protein Families

By grouping proteins into specific families, the functional analysis can be enhanced greatly. Different mathematical algorithms have been developed for such a classification such as applying hidden Markov models, position-specific matrices, and support vector

Table 8 Protein structure prediction tools

Name	Description	URL
MODELLER	Homology modeling	https://modbase.compbio.ucsf.edu/modweb/
SWISS-MODEL	Homology modeling	https://swissmodel.expasy.org/
WHAT IF	Homology modeling	https://web.archive.org/web/20130618170139/ http://swift.cmbi.kun.nl/whatif/
Yasara	Homology modeling	http://www.yasara.org/
RaptorX	Homology modeling	http://raptorx.uchicago.edu
IntFOLD	Homology modeling	http://www.reading.ac.uk/bioinf/IntFOLD/
FoldX	Homology modeling	http://foldxsuite.crg.eu/
ROBETTA	Homology modeling and ab initio	http://robetta.bakerlab.org/
I-TASSER	Fold recognition and ab initio	https://zhanglab.ccmb.med.umich.edu/I-TASSER/
NovaFold	Fold recognition and ab initio	http://www.dnastar.com/t-products-NovaFold.aspx
SPARKS-X	Fold recognition	http://sparks-lab.org/yueyang/server/SPARKS-X/
EVfold	Ab initio	http://evfold.org/evfold-web/evfold.do
FALCON	Ab initio	http://protein.ict.ac.cn/FALCON/
QUARK	Ab initio	https://zhanglab.ccmb.med.umich.edu/QUARK/

Table 9 Protein 3D structure alignment and comparison

Name	Description	URL
Dali server	Structure comparisons	http://ekhidna2.biocenter.helsinki.fi/dali/
iSARST	The integrated service of structural similarity search aided by ramachandran sequential transformation	http://140.113.15.73/~lab/iSARST/srv/
MATRAS	Comparing protein structures using Matras	http://strcomp.protein.osaka-u.ac.jp/matras/
RAPIDO	3D alignment of crystal structures of different protein molecules in the presence of conformational change	http://webapps.embl-hamburg.de/rapido/
deconSTRUCT	Protein structure database search	http://epsf.bmad.bii.a-star.edu.sg/struct_server.html
YAKUNET	Fast scan of protein structure databases	http://wwwabi.snv.jussieu.fr/Yakusa/
SALAMI	Structural alignment for proteins	http://flensburg.zbh.uni-hamburg.de/~wurst/salami/
fastSCOP	For recognizing protein structural domains and SCOP superfamilies	http://fastscop.life.nctu.edu.tw/
Cloud4Psi	Uses a combination of three algorithms: jCE, jFATCAT-rigid, and jFATCAT-flexible	http://cloud4psi.cloudapp.net/
3D-Blast	Uses the Hex docking spherical polar Fourier (SPF) correlation technique	http://threedblast.loria.fr/

Table 10 Protein families, motifs, and domains databases

Name	Description	URL
PIRSF	Clustering of sequences and their evolutionary relationships	https://pir.georgetown.edu/pirwww/dbinfo/pirsf.shtml
COGs	Phylogenetic classification of proteins	http://www.ncbi.nlm.nih.gov/COG/
PANTHER	Classification based on family, function, biological process, and pathway	http://www.pantherdb.org/
ProtoNet	Hierarchical classification	http://www.protonet.cs.huji.ac.il/index.php
Pfam	Protein families, their alignments, and HMMs	http://www.pfam.sanger.ac.uk/
ProDom	Protein domain families	http://prodom.prabi.fr/prodom/current/html/home.php
CDD	Database of conserved domains	http://www.ncbi.nlm.nih.gov/sites/entrez?db=cdd
SMART	Simple modular architecture research tool	http://www.smart.embl.de/
PRINTS	Protein fingerprints	http://www.bioinf.manchester.ac.uk/dbbrowser/PRINTS/index.php
PROSITE	Protein families and domains	http://www.ca.expasy.org/prosite/
Interpro	Protein families and domains/sites prediction	http://www.ebi.ac.uk/interpro/
PFSCAN	Motif scanning	http://hits.isb-sib.ch/cgi-bin/PFSCAN?
Block	Multiply aligned ungapped segments corresponding to the most highly conserved regions of proteins	http://blocks.fhcrc.org/blocks/blocks_search.html

machines (Finn et al., 2015). A new protein sequence can be compared within a database of protein families using such mathematical representations. Finding the protein family of a sequence instead of finding individually similar sequences by searching in sequence databases can provide more knowledge about the target sequence (Table 10).

3.5 Interactions, Pathways, and System Biology

Proteins, in particular intracellular proteins, usually do not act individually and interact with other proteins in form of protein-protein interaction networks and pathways to function. Therefore, to fully understand the function of each protein, we need also study its interaction to other proteins, its role in pathways, and its subcellular location. Cell-wide protein systems and protein interaction maps are the subject of modern systems biology studies (see Tables 11–13).

Table 11 Molecular interaction databases

Name	Description	URL
DIP	Database of interacting proteins	http://dip.doe-mbi.ucla.edu/
Reactome		http://www.reactome.org/
PIMRider	Protein interaction maps	http://pim.hybrigenics.com/ pimriderext/common
BOND	Biomolecular object network databank	http://bond. unleashedinformatics.com/
OPHID	Online predicted human interaction database	http://128.100.65.8/ophidv2. 201/index.jsp
MINT	Molecular interactions database	http://mint.bio.uniroma2.it/ mint/Welcome.do
HPRD	Human protein reference database	http://www.hprd.org/
GRID	The general repository for interaction datasets	http://www.thebiogrid.org/
IntAct	Database and toolkit for the storage, presentation, and analysis of protein interactions	http://www.ebi.ac.uk/intact/ site/index.jsf
HPID	Human protein interaction database	http://wilab.inha.ac.kr/hpid/
AANT		http://aant.icmb.utexas.edu/
MIPS	Yeast protein-protein interaction data	http://mips.gsf.de/proj/ppi/
HAPPI	Human annotated and predicted protein interaction database	http://bio.informatics.iupui. edu/HAPPI/
IntNetDB	Predicted protein-protein interactions	http://hanlab.genetics.ac.cn/ IntNetDB.htm
MiMI	Molecular interaction data	http://mimi.ncibi.org/MiMI/ home.jsp
UniHi	Unified human interactome	http://theoderich.fb3.mdc-berlin.de:8080/unihi/home
CPDB	Allows searching, visualizing and retrieving of integrated interaction data	http://consensuspathdb.org/
BID	Binding interface database	http://tsailab.org/BID/index. php
KDBI	Database of kinetic data of biomolecular interactions	http://xin.cz3.nus.edu.sg/ group/kdbi/kdbi.asp
DOMINO	Domain peptide interactions database	http://mint.bio.uniroma2.it/ domino/search/ searchWelcome.do
MPPI	Mammalian protein-protein interaction database	http://mips.gsf.de/proj/ppi/
PDZBase	Manually curated protein-protein interaction database developed specifically for interactions involving PDZ domains	http://icb.med.cornell.edu/ services/pdz/start
PepCyber	Database of human protein-protein interactions mediated by phosphoprotein-binding domains (PPBDs)	http://pepcyber.umn.edu/ PPEP/
PRIME	Protein interactions and molecular information database	http://prime.ontology.ims.u-tokyo.ac.jp:8081/
SPIN-PP	Surface properties of interfaces—protein protein interfaces	http://trantor.bioc.columbia. edu/cgi-bin/SPIN/
UniHI	Unified human interactome	http://theoderich.fb3.mdc-berlin.de:8080/unihi/home

Table 12 Pathway databases

Name	Description	URL
Kyoto Encyclopedia of Genes and Genomes (KEGG)	Kyoto encyclopedia of genes and genomes	http://www.genome.jp/ kegg/pathway.html
BioCyc	Literature and computationally derived pathways	http://www.biocyc.org/
MetaCyc	Metabolic pathways and enzymes	http://www.metacyc.org/
MiMI	Michigan molecular interactions	http://www.mimitest.ncibi. org/MimiWeb/mainpage. jsp
DAVID	The database for annotation, visualization, and integrated discovery	http://david.abcc.ncifcrf.gov/
PID	Pathway interaction database	http://pid.nci.nih.gov/
HAPPI	Human annotated and predicted protein interaction database	http://bio.informatics.iupui. edu/HAPPI
BioCarta	Pathway database	http://www.biocarta.com/
CSNDB	Cell signaling networks database	http://www.chem.ac.ru/ Chemistry/Databases/ CSNDB.en.html
EMP	Enzymes and metabolic pathways	http://www.empproject. com/about/
SPAD	Signaling pathway database	http://www.grt.kyushu-u.ac. jp/spad/

3.6 Posttranslational Modifications

During the protein biosynthesis, especially in relation to important cell signaling components, different posttranslational modifications (PTMs) may happen through enzymatic modifications. Either amino acids within the protein chain or at its N- or C-terminals can be sites of PTMs, thus expanding chemical capability of the 20 standard amino acids by introducing new functional groups to them (Khoury et al., 2011). Phosphorylation is the most common place PTM that acts as a regulating mechanism for enzymes' activity. Glycosylation is another PTM that, besides regulatory functions, can improve the protein stability and lipidation, making it more compatible with the cell membrane. Other forms of PTM consist of disulfide bond formation, peptide bond cleavage, and carbonylation (Dalle-Donne et al., 2006; Grimsrud et al., 2008). For a comprehensive list of PTMs, look at http://www.uniprot.org/docs/ptmlist. Similar to other features of proteins, the PTM sites can also be predicted using in silico methods in comparison with existing experimentally reported PTMs. Table 14 lists some common PTM predictor tools.

Table 13 Interaction and pathway tools

Name	Description	URL
IPA	Pathway and functional annotation	http://www.ingenuity.com/
Cytoscape	An open source platform for complex network analysis and visualization	http://www.cytoscape.org/
BINGO	Biological network gene ontology tool	http://www.psb.ugent.be/cbd/papers/BiNGO/Home.html
GATHER	Gene annotation tool to help explain relationships	http://gather.genome.duke.edu
SCOPPI	Structural classification of protein-protein interfaces	http://www.scoppi.org/
ADVICE	Automated detection and validation of interaction by coevolution	http://advice.i2r.a-star.edu.sg/
BioLayout Java	Automatic graph layout algorithm for similarity and network visualization	http://cgg.ebi.ac.uk/services/biolayout/
eFsite	Electrostatic surface of Functional-SITE	http://ef-site.hgc.jp/eF-site/
Expression Profiler	Explores protein interaction data using expression data	http://ep.ebi.ac.uk/EP/PPI/
InterPreTS	Protein interactionprediction through tertiary structure	http://speedy.embl-heidelberg.de/people/patrick/interprets/index.html
InterViewer	Visualization of large-scale protein interaction networks	http://interviewer.inha.ac.kr/
iPPI	Infers protein-protein interactions through homology search	http://www.bioinfo.cu/iPPI.html
IPPRED	Infers interactions through homology search	http://cbi.labri.fr/outils/ippred/IS_part_simple.php
iSPOT	Prediction of protein-protein interactions mediated by families of peptide recognition modules	http://cbm.bio.uniroma2.it/ispot/
NetAlign	Comparison of protein interaction networks	http://www1.ustc.edu.cn/lab/pcrystal/NetAlign/
PathBLAST	Comparing protein interaction networks	http://www.pathblast.org/
PIVOT	Protein interactions visualization tool	http://www.cs.tau.ac.il/~rshamir/pivot/
SPIN-PP	Surface properties of interfaces—protein protein interfaces	http://trantor.bioc.columbia.edu/cgi-bin/SPIN/
STRING	Search tool for the retrieval of interacting genes/proteins	http://string.embl.de/
PROPER	Global protein interaction network alignment through percolation matching	http://proper.epfl.ch

Table 14 Protein posttranslational modification prediction

Name	Description	URL
FindMod	Predicts the PTM for a given protein sequence	http://us.expasy.org/tools/findmod/
MITOProt	Finds the N-terminal region of a protein that can support a mitochondrial targeting sequence and the cleavage site as well	http://www.mips.biochem.mpg.de/cgi-bin/proj/medgen/mitofilter
NetNGlyc	For a human protein, it uses artificial neural network it predicts the N-glycosylation sites	http://www.cbs.dtu.dk/services/NetNGlyc/
NetOGlyc	For mammalian proteins, it predicts mucin typeGalNAc O-glycosylation sites	http://www.cbs.dtu.dk/services/NetOGlyc/
NetPhos	For eukaryotic proteins, it predicts threonine, serine, and tyrosine phosphorylation sites	http://www.cbs.dtu.dk/services/NetPhos/
SignalP	Predicts signal peptide cleavage sites' locations for a given protein sequence	http://www.cbs.dtu.dk/services/SignalP/
Sulfinator	In a protein sequence, it predicts the tyrosine sulfation sites	http://us.expasy.org/tools/sulfinator/
TargetP	For eukaryotic protein sequences, it predicts their subcellular location	http://www.cbs.dtu.dk/services/TargetP/

4. CONCLUSION

The last decades have seen an explosion of biological data with advances in technology, especially DNA sequencing and high-throughput analysis tools. The available data in different databases can be used to discover the big map of the life. Moreover, data analysis tools can enable the creation and training of different models that will eventually help to develop life science technologies. With the advent of new computational approaches and experimental techniques, many challenges still exist, and hence many opportunities are presented, toward further development of modern proteomics.

REFERENCES

Alaoui-Jamali, M.A., Xu, Y.-J., 2006. Proteomic technology for biomarker profiling in cancer: an update. J. Zhejiang Univ. Sci. B 7, 411–420.

Altschul, S.F., Gish, W., Miller, W., Myers, E.W., Lipman, D.J., 1990. Basic local alignment search tool. J. Mol. Biol. 215, 403–410.

Altschul, S.F., Madden, T.L., Schäffer, A.A., Zhang, J., Zhang, Z., Miller, W., Lipman, D.J., 1997. Gapped BLAST and PSI-BLAST: a new generation of protein database search programs. Nucleic Acids Res. 25, 3389–3402.

Amico, M., Finelli, M., Rossi, I., Zauli, A., Elofsson, A., Viklund, H., von Heijne, G., Jones, D., Krogh, A., Fariselli, P., 2006. PONGO: a web server for multiple predictions of all-alpha transmembrane proteins. Nucleic Acids Res. 34, W169–W172.

Anderson, N.L., Anderson, N.G., 1998. Proteome and proteomics: new technologies, new concepts, and new words. Electrophoresis 19, 1853–1861.

Anderson, N.L., Anderson, N.G., 2002. The human plasma proteome history, character, and diagnostic prospects. Mol. Cell. Proteomics 1, 845–867.

Baxevanis, A.D., Ouellette, B.F., 2004. Bioinformatics: A Practical Guide to the Analysis of Genes and Proteins. John Wiley & Sons, New York.

Björck, L., Kronvall, G., 1984. Purification and some properties of streptococcal protein G, a novel IgG-binding reagent. J. Immunol. 133, 969–974.

Blackstock, W.P., Weir, M.P., 1999. Proteomics: quantitative and physical mapping of cellular proteins. Trends Biotechnol. 17, 121–127.

Blom, N., Sicheritz-Pontén, T., Gupta, R., Gammeltoft, S., Brunak, S., 2004. Prediction of post-translational glycosylation and phosphorylation of proteins from the amino acid sequence. Proteomics 4, 1633–1649.

Blonder, J., Goshe, M.B., Moore, R.J., Pasa-Tolic, L., Masselon, C.D., Lipton, M.S., Smith, R.D., 2002. Enrichment of integral membrane proteins for proteomic analysis using liquid chromatography-tandem mass spectrometry. J. Proteome Res. 1, 351–360.

Bujnicki, J.M., 2008. Prediction of Protein Structures, Functions, and Interactions. John Wiley & Sons, West Sussex, UK.

Castronovo, V., Waltregny, D., Kischel, P., Roesli, C., Elia, G., Rybak, J.-N., Neri, D., 2006. A chemical proteomics approach for the identification of accessible antigens expressed in human kidney cancer. Mol. Cell. Proteomics 5, 2083–2091.

Chandrasekhar, K., Dileep, A., Lebonah, D.E., 2014. A short review on proteomics and its applications. Int. Lett. Nat. Sci 12, 77–84.

Dalle-Donne, I., Aldini, G., Carini, M., Colombo, R., Rossi, R., Milzani, A., 2006. Protein carbonylation, cellular dysfunction, and disease progression. J. Cell. Mol. Med. 10, 389–406.

Devos, D., Valencia, A., 2000. Practical limits of function prediction. Proteins: Struct., Funct., Bioinf. 41, 98–107.

Emanuelsson, O., Brunak, S., Von Heijne, G., Nielsen, H., 2007. Locating proteins in the cell using TargetP, SignalP and related tools. Nat. Protoc. 2, 953.

Finn, R.D., Coggill, P., Eberhardt, R.Y., Eddy, S.R., Mistry, J., Mitchell, A.L., Potter, S.C., Punta, M., Qureshi, M., Sangrador-Vegas, A., 2015. The Pfam protein families database: towards a more sustainable future. Nucleic Acids Res. 44, D279–D285.

Godovac-Zimmermann, J., Brown, L.R., 2001. Perspectives for mass spectrometry and functional proteomics. Mass Spectrom. Rev. 20, 1–57.

Govorukhina, N., Keizer-Gunnink, A., Van der Zee, A., De Jong, S., De Bruijn, H., Bischoff, R., 2003. Sample preparation of human serum for the analysis of tumor markers: comparison of different approaches for albumin and γ-globulin depletion. J. Chromatogr. A 1009, 171–178.

Grimsrud, P.A., Xie, H., Griffin, T.J., Bernlohr, D.A., 2008. Oxidative stress and covalent modification of protein with bioactive aldehydes. J. Biol. Chem. 283, 21837–21841.

Guex, N., Peitsch, M.C., 1997. SWISS-MODEL and the Swiss-Pdb viewer: an environment for comparative protein modeling. Electrophoresis 18, 2714–2723.

Gygi, S.P., Rist, B., Gerber, S.A., Turecek, F., Gelb, M.H., Aebersold, R., 1999. Quantitative analysis of complex protein mixtures using isotope-coded affinity tags. Nat. Biotechnol 17, 994–999.

Han, D.K., Eng, J., Zhou, H., Aebersold, R., 2001. Quantitative profiling of differentiation-induced microsomal proteins using isotope-coded affinity tags and mass spectrometry. Nat. Biotechnol. 19, 946–951.

Hawkins, T., Luban, S., Kihara, D., 2006. Enhanced automated function prediction using distantly related sequences and contextual association by PFP. Protein Sci. 15, 1550–1556.

Horton, P., Park, K.-J., Obayashi, T., Fujita, N., Harada, H., Adams-Collier, C., Nakai, K., 2007. WoLF PSORT: protein localization predictor. Nucleic Acids Res. 35, W585–W587.

Jollès, P., Jörnvall, H., 2013. Proteomics in Functional Genomics: Protein Structure Analysis. Birkhäuser, Basel.

Julenius, K., 2007. NetCGlyc 1.0: prediction of mammalian C-mannosylation sites. Glycobiology 17, 868–876.

Julenius, K., Mølgaard, A., Gupta, R., Brunak, S., 2005. Prediction, conservation analysis, and structural characterization of mammalian mucin-type O-glycosylation sites. Glycobiology 15, 153–164.

Juncker, A.S., Jensen, L.J., Pierleoni, A., Bernsel, A., Tress, M.L., Bork, P., Von Heijne, G., Valencia, A., Ouzounis, C.A., Casadio, R., 2009. Sequence-based feature prediction and annotation of proteins. Genome Biol. 10, 206.

Kent, W.J., 2002. BLAT—the BLAST-like alignment tool. Genome Res. 12, 656–664.

Khoury, G.A., Baliban, R.C., Floudas, C.A., 2011. Proteome-wide post-translational modification statistics: frequency analysis and curation of the swiss-prot database. Sci. Rep. 1, 90.

Kihara, D., 2016. Protein Structure Prediction. Springer, New York, pp. 1–15.

Klose, J., Nock, C., Herrmann, M., Stühler, K., Marcus, K., Blüggel, M., Krause, E., Schalkwyk, L.C., Rastan, S., Brown, S.D., 2002. Genetic analysis of the mouse brain proteome. Nat. Genet. 30, 385.

Lei, T., He, Q.-Y., Wang, Y.-L., Si, L.-S., Chiu, J.-F., 2008. Heparin chromatography to deplete high-abundance proteins for serum proteomics. Clin. Chim. Acta 388, 173–178.

Linding, R., Jensen, L.J., Ostheimer, G.J., van Vugt, M.A., Jørgensen, C., Miron, I.M., Diella, F., Colwill, K., Taylor, L., Elder, K., 2007. Systematic discovery of in vivo phosphorylation networks. Cell 129, 1415–1426.

López, G., Valencia, A., Tress, M.L., 2007. Firestar—prediction of functionally important residues using structural templates and alignment reliability. Nucleic Acids Res. 35, W573–W577.

Melen, K., Krogh, A., von Heijne, G., 2003. Reliability measures for membrane protein topology prediction algorithms. J. Mol. Biol. 327, 735–744.

Melton, L., 2004. Protein arrays: proteomics in multiplex. Nature 429, 101–107.

Miller, M.L., Jensen, L.J., Diella, F., Jørgensen, C., Tinti, M., Li, L., Hsiung, M., Parker, S.A., Bordeaux, J., Sicheritz-Ponten, T., 2008. Linear motif atlas for phosphorylation-dependent signaling. Sci. Signal. 1, ra2.

Mueller, L.N., Rinner, O., Schmidt, A., Letarte, S., Bodenmiller, B., Brusniak, M.Y., Vitek, O., Aebersold, R., Müller, M., 2007. SuperHirn—a novel tool for high resolution LC-MS-based peptide/protein profiling. Proteomics 7, 3470–3480.

Nair, R., Rost, B., 2002. Sequence conserved for subcellular localization. Protein Sci. 11, 2836–2847.

Nedelkov, D., Nelson, R.W., 2006. New and Emerging Proteomic Techniques. Humana Press Inc, New Jersey, pp. 159–175.

Obenauer, J.C., Cantley, L.C., Yaffe, M.B., 2003. Scansite 2.0: proteome-wide prediction of cell signaling interactions using short sequence motifs. Nucleic Acids Res. 31, 3635–3641.

Ong, S.-E., Blagoev, B., Kratchmarova, I., Kristensen, D.B., Steen, H., Pandey, A., Mann, M., 2002. Stable isotope labeling by amino acids in cell culture, SILAC, as a simple and accurate approach to expression proteomics. Mol. Cell. Proteomics 1, 376–386.

Pieper, U., Eswar, N., Stuart, A.C., Ilyin, V.A., Sali, A., 2002. MODBASE, a database of annotated comparative protein structure models. Nucleic Acids Res. 30, 255–259.

Pierleoni, A., Martelli, P.L., Fariselli, P., Casadio, R., 2006. BaCelLo: a balanced subcellular localization predictor. Bioinformatics 22, e408–e416.

Porter, C.T., Bartlett, G.J., Thornton, J.M., 2004. The catalytic site atlas: a resource of catalytic sites and residues identified in enzymes using structural data. Nucleic Acids Res. 32, D129–D133.

Ross, P.L., Huang, Y.N., Marchese, J.N., Williamson, B., Parker, K., Hattan, S., Khainovski, N., Pillai, S., Dey, S., Daniels, S., 2004. Multiplexed protein quantitation in Saccharomyces cerevisiae using amine-reactive isobaric tagging reagents. Mol. Cell. Proteomics 3, 1154–1169.

Rothemund, D.L., Locke, V.L., Liew, A., Thomas, T.M., Wasinger, V., Rylatt, D.B., 2003. Depletion of the highly abundant protein albumin from human plasma using the Gradiflow. Proteomics 3, 279–287.

Samaj, J., Thelen, J.J., 2007. Plant Proteomics. Springer-Verlag, pp. 1–13.

Sato, A.K., Sexton, D.J., Morganelli, L.A., Cohen, E.H., Wu, Q.L., Conley, G.P., Streltsova, Z., Lee, S.W., Devlin, M., DeOliveira, D.B., 2002. Development of mammalian serum albumin affinity purification media by peptide phage display. Biotechnol. Prog. 18, 182–192.

Scheurer, S.B., Rybak, J.N., Roesli, C., Brunisholz, R.A., Potthast, F., Schlapbach, R., Neri, D., Elia, G., 2005. Identification and relative quantification of membrane proteins by surface biotinylation and two-dimensional peptide mapping. Proteomics 5, 2718–2728.

Schulz, T.C., Swistowska, A.M., Liu, Y., Swistowski, A., Palmarini, G., Brimble, S.N., Sherrer, E., Robins, A.J., Rao, M.S., Zeng, X., 2007. A large-scale proteomic analysis of human embryonic stem cells. BMC Genomics 8, 478.

Seitz, P., Modarres, H.P., Borgeaud, S., Bulushev, R.D., Steinbock, L.J., Radenovic, A., Dal Peraro, M., Blokesch, M., 2014. ComEA is essential for the transfer of external DNA into the periplasm in naturally transformable Vibrio cholerae cells. PLoS Genet. 10, e1004066.

Smith, J.C., Lambert, J.-P., Elisma, F., Figeys, D., 2007. Proteomics in 2005/2006: developments, applications and challenges. Anal. Chem. 79, 4325–4344.

Stahl-Zeng, J., Lange, V., Ossola, R., Eckhardt, K., Krek, W., Aebersold, R., Domon, B., 2007. High sensitivity detection of plasma proteins by multiple reaction monitoring of N-glycosites. Mol. Cell. Proteomics 6, 1809–1817.

Team, M.G.C.P., 2002. Generation and initial analysis of more than 15,000 full-length human and mouse cDNA sequences. Proc. Natl. Acad. Sci. USA 99, 16899–16903.

Travis, J., Bowen, J., Tewksbury, D., Johnson, D., Pannell, R., 1976. Isolation of albumin from whole human plasma and fractionation of albumin-depleted plasma. Biochem. J. 157, 301–306.

Van den Bergh, G., Arckens, L., 2005. Recent advances in 2D electrophoresis: an array of possibilities. Expert Rev. Proteomics 2, 243–252.

Viklund, H., Elofsson, A., 2004. Best α-helical transmembrane protein topology predictions are achieved using hidden Markov models and evolutionary information. Protein Sci. 13, 1908–1917.

Washburn, M.P., Wolters, D., Yates III, J.R., 2001. Large-scale analysis of the yeast proteome by multidimensional protein identification technology. Nat. Biotechnol. 19, 242.

Wass, M.N., Sternberg, M.J., 2008. ConFunc—functional annotation in the twilight zone. Bioinformatics 24, 798–806.

Wilkins, M.R., Sanchez, J.-C., Gooley, A.A., Appel, R.D., Humphery-Smith, I., Hochstrasser, D.F., Williams, K.L., 1996. Progress with proteome projects: why all proteins expressed by a genome should be identified and how to do it. Biotechnol. Genet. Eng. Rev. 13, 19–50.

Wilkins, M.R., Gasteiger, E., Gooley, A.A., Herbert, B.R., Molloy, M.P., Binz, P.-A., Ou, K., Sanchez, J.-C., Bairoch, A., Williams, K.L., 1999. High-throughput mass spectrometric discovery of protein post-translational modifications. J. Mol. Biol. 289, 645–657 (11 Edited by Huber, R.).

Wolters, D.A., Washburn, M.P., Yates, J.R., 2001. An automated multidimensional protein identification technology for shotgun proteomics. Anal. Chem. 73, 5683–5690.

Xu, Y., Xu, D., Liang, J., 2007. Computational Methods for Protein Structure Prediction and Modeling. Springer, New York.

Zhang, H., Yan, W., Aebersold, R., 2004. Chemical probes and tandem mass spectrometry: a strategy for the quantitative analysis of proteomes and subproteomes. Curr. Opin. Chem. Biol. 8, 66–75.

Zuo, X., Speicher, D.W., 2000. A method for global analysis of complex proteomes using sample prefractionation by solution isoelectrofocusing prior to two-dimensional electrophoresis. Anal. Biochem. 284, 266–278.

CHAPTER 3

Understanding Specialized Ribosomal Protein Functions and Associated Ribosomopathies by Navigating Across Sequence, Literature, and Phenotype Information Resources

K.A. Kyritsis*, L. Angelis[†], Christos Ouzounis[‡], Ioannis Vizirianakis*

[*]Laboratory of Pharmacology, School of Pharmacy, Aristotle University of Thessaloniki, Thessaloniki, Greece
[†]Department of Informatics, Aristotle University of Thessaloniki, Thessaloniki, Greece
[‡]Biological Computation & Process Laboratory, Chemical Process & Energy Resources Institute, Centre for Research & Technology Hellas, Thermi, Greece

1. INTRODUCTION

Ribosome biogenesis is a highly coordinated cellular process that involves a multitude of macromolecular components, leading to the stoichiometric assembly of the ribosome. In human, 4 rRNAs and 80 ribosomal proteins (RPs) are expressed, processed, and assembled into the large (60S) and the small (40S) ribosomal subunits, each of which possesses specific functions for the translation of mRNA into protein (Kressler et al., 2010; Lafontaine, 2015). While the textbook depiction of translation is typically represented by a universal scheme identical across cells and tissues, recent studies have challenged this notion. For example, it has been demonstrated that RPs possess tissue-specific functional roles, beyond the constitutive catalysis of mRNA translation and the structural integrity of ribosome (Xue and Barna, 2012). Notably, mutations in RP genes, which reduce their expression (haploinsufficiency), cause the development of rare congenital diseases and are associated with cancer predisposition (Sulima et al., 2014; De Keersmaecker et al., 2015; Yelick and Trainor, 2015). Herein, we provide a cursory critical review of the medical literature for reported, multifunctional roles of RPs and their association with disease phenotypes. Furthermore, we deploy simple metrics for orthologous RPs to measure their sequence conservation, and assess how these are reflected in the relevant biomedical literature. Our results constitute a first step toward a deeper understanding of the functionality of RPs and how its disruption results in a range of pathological conditions.

Leveraging Biomedical and Healthcare Data
https://doi.org/10.1016/B978-0-12-809556-0.00003-4

2. RPS AND DISEASES

2.1 Ribosomopathies

Haploinsufficiency of certain RP genes may lead to perturbation of ribosome assembly and translation efficiency, resulting in the appearance of rare clinical syndromes, known as ribosomopathies. Despite the fact that most RPs are ubiquitously expressed across all tissues, ribosomopathies present highly specific symptoms in patients, including hematological disorders, bone marrow aplasia, and cancer predisposition (Sulima et al., 2014; De Keersmaecker et al., 2015; Yelick and Trainor, 2015). For example, RPS14 haploinsufficiency plays a critical role in the 5q deletion (5q-) syndrome, a subtype of myelodysplastic syndromes. Overexpression of RPS14 leads to the rescue of the defective erythroid differentiation phenotype in 5q- (Ebert et al., 2008). Haploinsufficiency of RPS14 was shown to result in anemia via both p53-dependent and p53-independent mechanisms (Nakhoul et al., 2014). Another ribosomopathy is Diamond-Blackfan Anemia (DBA), a congenital hypoplastic anemia caused by heterozygous mutations in several RP genes, such as RPS10, RPS26, RPS24, RPS17, RPS7, RPL35a, RPL11, RPL5, RPL26, RPL15, and RPS19, which account for over 50% of all DBA cases (Dahl et al., 1999; Boria et al., 2010; Ruggero and Shimamura, 2014). The latter disease exhibits clinical heterogeneity that is partially dependent by the RP genes carrying the causal mutations, and is primarily characterized by red blood cell aplasia, macrocytic anemia, and cancer susceptibility (Ellis and Lipton, 2008; Gazda et al., 2008).

To date, only a fraction of the entire spectrum of molecular pathogenesis pathways underlying ribosomopathies has been adequately characterized. Recent studies suggest the abnormal activation of p53, resulting from perturbations in ribosome biogenesis, as one of the leading causes for pathological manifestations in ribosomopathies (Golomb et al., 2014). Inhibition of p53 rescues the erythroid defective phenotype in DBA animal models (Danilova et al., 2008; McGowan et al., 2008), while knockdown of RPS14 and RPS19 in human hematopoietic progenitor cells selectively activates p53 thus impairing erythropoiesis (Dutt et al., 2011). Moreover, nuclear p53 was found accumulated in erythroid progenitor cells, received from bone marrow biopsy samples of 5q- and DBA patients (Dutt et al., 2011), and lenalidomide, a commonly prescribed drug for DBA and 5q- syndrome, was recently shown to act through promoting p53 degradation (Wei et al., 2013a).

Interestingly, in a small number of DBA patients, lacking pathological mutations in RP genes, next-generation sequencing revealed rare mutations in the GATA1 gene. These mutations resulted in impaired production of the full-length form of GATA1, known to be specifically upregulated during erythroid differentiation (Sankaran et al., 2012; Ludwig et al., 2014). Furthermore, it was reported that knockdown of RPS19, RPL11, RPL5, or RPS24 leads to reduced protein levels of GATA1 in CD34+ derived erythroid progenitors and in K562 erythroid cell line cultures in vitro, without affecting

GATA1 mRNA levels (Ludwig et al., 2014). Similar observations have been made for Hsp70 cochaperone BAG1 and the RNA-binding protein CSDE1: knockdown of RPS19 or RPL11 in mouse erythroblasts resulted in reduced protein levels of BAG1 and CSDE1, while their mRNA levels remained unchanged (Horos et al., 2012). Reduced protein but not mRNA levels were also observed in erythroblasts obtained from DBA patients, compared to erythroblasts from healthy donors. Importantly, both BAG1 and CSDE1 carry internal ribosome entry site (IRES) elements in the 5′ untranslated region (UTR) of their mRNAs, while decrease in the expression of either gene results in impairment of normal erythropoiesis (Horos et al., 2012).

These important findings suggest that the translational efficiency of hematopoietic factor mRNAs, like GATA1, BAG1 and CSDE1, depends on RP-mediated regulation, whose disruption can lead to the erythroid defective phenotype that characterizes ribosomopathies. Conversely, as proposed recently, specific hematopoietic factors, such as GATA1 and PU.1, are involved in the transcriptional regulation of RP gene expression upon differentiation of myeloid progenitor cells to erythroid maturation (Amanatiadou et al., 2015). This overview provides a quick tour of the complex role of RPs and associated genes in ribosomal dysfunction resulting in ribosomopathies.

2.2 Cancer

Increased levels of ribosome biogenesis constitute a specific trait for cancer development, as the increase of protein synthesis levels is a necessary element for cancer cell growth and proliferation (Golomb et al., 2014; Dez and Tollervey, 2004). It is also worth noting that patients of several ribosomopathies, such as DBA, display a high risk for cancer in older age (Vlachos et al., 2012).

In model organism experiments, increased frequency of cancer incidents, primarily of malignant peripheral nerve sheath tumors, was reported for eleven zebrafish RP-associated models, each carrying a heterozygous mutation for a different RP gene (Amsterdam et al., 2004). Furthermore, the RPL22 locus was found monoallelically deleted with frequency of ~ 10%, in a cohort of primary T-acute lymphoblastic leukemia (T-ALL) cases (Rao et al., 2012). The same study reported that heterozygous inactivation of RPL22 in a T-ALL mouse model enhanced the development of the disease (Rao et al., 2012), possibly through the induction of the tumorigenic factor Lin28B (Viswanathan et al., 2009). Mutations in RPL22 were also detected in a computational analysis of mutation frequencies across 12 cancer types (Kandoth et al., 2013). Finally, mutations affecting RPL5 and RPL10 proteins were detected in a subset of pediatric T-ALL cases (De Keersmaecker et al., 2013).

Recent in silico analysis of RNA-seq data from The Cancer Genome Atlas (TCGA) project has shown a systematic reduction of expression levels for specific RPs with tumor-suppressive activity, such as RPL5 and RPL11 populations (Guimaraes and

Zavolan, 2016). Conversely, RPL36A, RPS2, and RPL39L proteins, generally associated with inducing cell proliferation (Kim et al., 2004; Wang et al., 2009; Wong et al., 2014), display an increased level of expression. Furthermore, the transcriptional levels of RPL39L, RPL26L1, and RPL21 have yielded high accuracy in predicting the survival of patients with breast cancer (Guimaraes and Zavolan, 2016). As with ribosomopathies, the involvement of RPs in cancer development is associated with a complex interplay between variation and misregulated expression.

3. SPECIALIZED FUNCTIONS OF RPS

The unexpected specificity of the phenotypes arising from disruptions in RP gene expression strongly implies that these proteins possess additional cellular roles. Indeed, many recent studies have revealed the involvement of RPs in the regulation of the translational efficiency of distinct mRNAs (Xue and Barna, 2012). These regulatory functions depend on various factors, including the presence of specific cis-regulatory elements within mRNAs, posttranslational modifications (PTMs) of RPs, as well as the expression levels and the stoichiometry of RPs and their paralogs (Xue and Barna, 2012).

3.1 RP-Mediated Translational Control of Distinct mRNAs, Through Interaction With IRES cis-Regulatory Elements

RPs regulate the translation of distinct mRNAs by interacting with cis–regulatory elements located in them, such as IRES elements (Xue and Barna, 2015). IRES constitute cis-regulatory elements located at the 5′ UTR, which serves in recruiting the 40S small ribosomal subunit and initiating the translation of the mRNA in the absence of 5′ cap (Xue and Barna, 2012).

A well-studied example of RP control of mRNA translation through IRES is RPL38 and Hox genes. RPL38 was found to be essential for the accurate gene expression of a subset of Hox genes during murine embryogenesis (Kondrashov et al., 2011). These particular Hox genes carry IRES elements as well as a TIE cis-element (Translational Inhibitory Element), preventing the addition of 5′ cap in their mRNAs. Thus, their accurate expression, which is required for normal development, is regulated by RPL38 through its interaction with the IRES elements of Hox mRNAs (Xue et al., 2015). RPL38 mutations in murine embryos cause pathological, tissue-specific phenotypes, which include perturbations in the formation of the axial skeleton (Kondrashov et al., 2011).

Generally, IRES are enriched in viral mRNAs and have been considered as a way for viruses to "hijack" the cellular machinery of translation (Sonenberg and Hinnebusch, 2009; Plank and Kieft, 2012). However, some well-characterized cellular genes with tumorigenic or tumor suppressive properties, including c-myc and p53, have been discovered to also carry IRES elements (Stoneley et al., 1998; Ray et al., 2006; Bellodi et al., 2010). Under certain conditions, such as mitosis, differentiation, and apoptosis, translation through the addition of the 7-methylguanylate cap at the 5′ end of mRNAs is partially inhibited. During this state, it has been shown that ~10% to 15% of all cellular mRNAs undergo translation through a cap-independent process, including IRES elements-mediated translation (Spriggs et al., 2008).

3.2 Posttranslational Modifications (PTMs) of RPs

Many RPs have been found to carry various posttranslational modifications (PTMs), including methylation, acetylation, phosphorylation, ubiquitylation, and glycosylation (Lee et al., 2002; Carroll et al., 2008; Odintsova et al., 2003; Yu et al., 2009; Krieg et al., 1988; Zeidan et al., 2010). As an example, the phosphorylation of RPS6 constitutes the most well-studied PTM in RPs. In higher eukaryotes, RPS6 is phosphorylated at its conserved carboxy-terminus, on five serine residues. RPS6 phosphorylation is induced by various stimuli and has been associated with physiological and pathological conditions (Meyuhas, 2015). Other tissue, developmental stage, and external conditions may determine the expression of PTM-catalyzing enzymes, increasing the potential of a highly diverse functionality for RPs.

3.3 RP Gene Expression

In mammals, a large-scale qPCR analysis was performed in 14 different tissues and cell types isolated from murine embryos, aiming at the understanding of gene expression for 72 RPs. Interestingly, a great level of heterogeneity was observed both for different RPs and various tissues, leading to the detection of RP clusters with similar expression patterns (Kondrashov et al., 2011). In a different study, qPCR was also employed to characterize the expression profile of 89 RPs in 22 adult human tissues. From the analyzed RPs, 21 were reported to exhibit highly heterogeneous expression. This distinct cluster of RPs was enriched in recently evolved RP paralogs while conservation analysis of promoter sequences revealed similar levels of conservation across all RPs (Wong et al., 2014). Furthermore, an extensive in silico analysis of RNA-seq data, originating from a variety of adult human tissues and primary cell populations (FANTOM Consortium and the RIKEN PMI and CLST (DGT) et al., 2014), was performed for the study of RP expression profiles. Of the 71 canonical, cytoplasmic RPs and the 19 paralog RP genes studied, a quarter exhibited tissue-specific expression patterns, especially in hematopoietic cellular populations. Notably, the list of tissue-specific genes was enriched in paralog RP genes (Guimaraes and Zavolan, 2016). Therefore, the heterogeneity of RP gene expression in different tissues and developmental stages supports their specialized functionality at least in human and mouse.

3.4 RP Paralogs

RP paralogs constitute a source of heterogeneity and specialized functionality for the ribosome. While in mammals most functional RPs are encoded by one gene, in other organisms, including *Saccharomyces cerevisiae* (budding yeast) (Kellis et al., 2004), *Drosophila* (Marygold et al., 2007), and *Arabidopsis thaliana* (Barakat et al., 2001), there is a varying number of gene copies for each RP. These variants diverge from their expected redundancy behavior as paralogs: in budding yeast, deletion of paralog genes results in unique phenotypes with altered metabolism, reproduction, and response to pathogens. Moreover, in higher eukaryotes like *Drosophila* and *Arabidopsis*, several RP paralogs display distinct expression patterns in different tissues and developmental stages, and are necessary for normal development (Xue and Barna, 2012).

3.5 Differential Subcellular Stoichiometry of RPs

Quantitative mass spectrometry analysis of monosome- and polysome-enriched cell extracts from budding yeast (*Saccharomyces cerevisiae*) and mouse embryonic stem cells (ESCs) revealed significant differences in the stoichiometry of a subset of RPs. Therefore, distinct subpopulations of ribosomes exist within the cell, differing in their composition of RPs and thus in their translational functionality (Slavov et al., 2015).

Differences in the stoichiometry of RPs between ribosomes appear to be conserved, change dynamically, and depend not only on the increasing number of ribosomes per mRNA, but also on the presence of external stimuli. These observations support the idea that differences in RPs stoichiometry could represent a mechanism for controlling levels of specific proteins by regulating the translational efficiency of their respective transcripts (Slavov et al., 2015). All the evidence strongly challenges the idea that RP function is universal and points toward the direction for research into the multiple roles of RP genes at various levels, locations, and conditions.

4. EXPLORING RP ROLES IN HEALTH AND DISEASE BY NAVIGATING BIOINFORMATICS RESOURCES

We have recently embarked into the analysis of human and mouse RPs using several bioinformatics approaches and tools, in order to explore the connection of RPs with human disease, their multiple cellular roles and their tissue-specific expression patterns, and other aspects of their structure and function.

4.1 RP Nomenclature

A new naming system for RPs (new nomenclature) was recently proposed to avoid issues arising from assigning identical names to RPs, which are not homologous and have significant structural differences (Ban et al., 2014).

In our study, RP gene names for both the old and the new nomenclature were retrieved from the site www.ribosomalproteins.com (Nicolas et al., 2016).

4.2 Comparison of Human and Mouse RPs Reveals a Range of Conservation Levels

Aiming at the investigation of RP conservation levels between human and mouse, the most widely used mammalian model for in vivo studies, we have performed a simple pairwise comparison of the 119 human RP protein sequences with their respective 85 mouse orthologue sequences. In our analysis, we have used the previous RP nomenclature to retrieve RefSeq entries, as protein sequences from the NCBI Protein database. Pairwise alignments to estimate identity levels were performed using NCBI's BLASTP software with default settings (Altschul et al., 1990).

Remarkably, and despite the fact that RPs are highly conserved, we were able to identify a varying degree of conservation at a range of 75% to 100% sequence identity, as well as multiple occurrences of gaps in the pairwise alignment of the ortholog pairs (see later).

✦ *Literature coverage of RPs reflects a varying degree of characterization*

Next, we have attempted to discover the extent at which each RP has been studied, by simple queries to the NCBI's PubMed database. In all, 82 RPs have been analyzed, including RPL27L and human RP paralogs RPS4Y1 and RPS4Y2, using the "reutils" R package. As the new RP nomenclature was proposed recently, we applied a composite query for PubMed database to capture all naming schemes. Specifically, we utilized

1. Gene names from the old RP nomenclature, which is used in almost all publications related to RPs.
2. Gene names from the new RP nomenclature but only for very recently published articles [specifically after the publication and establishment of www. ribosomalproteins.com, aiming at avoiding false positives (Nicolas et al., 2016)].

As an example, this is the query performed for RPS19 (now renamed as eS19), an RP directly associated with DBA:

RPS19[Title/Abstract] OR (("2016/07/01"[Date—Publication] : "3000"[Date—Publication]) AND eS19[Title/Abstract] AND ribosomal[Title/Abstract])

In this way, we aim at the highest possible coverage in the retrieval of all RP-related, PubMed articles while supporting the new RP nomenclature. Interestingly, the level of literature coverage for RPs ranges widely, with some RPs having been extensively studied as reflected by the number of returned abstract entries, while others are apparently less well characterized by targeted experimental analysis. In particular, RPS6 constitutes one of the well-studied RP with 468 PubMed articles, followed by RPS16 and RPS19 with 255 and 249 articles, respectively (Fig. 1). In contrast, RPL18a appears as the least studied RP with only 6 PubMed references (Fig. 1).

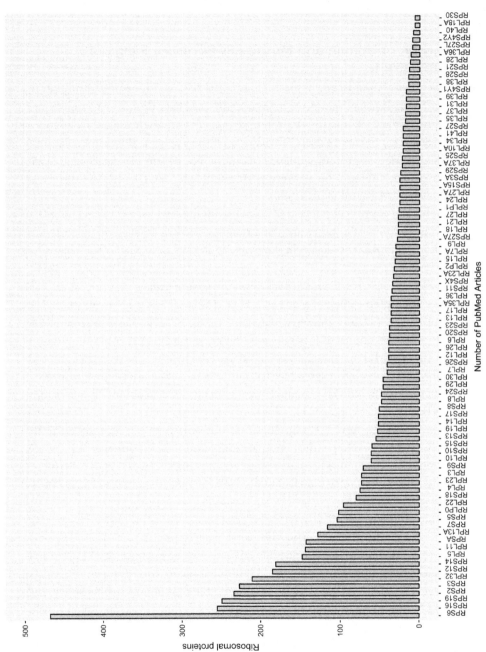

Fig. 1 Number of PubMed articles in which each RP identifier has been detected. All queries were performed in February 2017. The *bar plot was drawn using ggplot2 R package.*

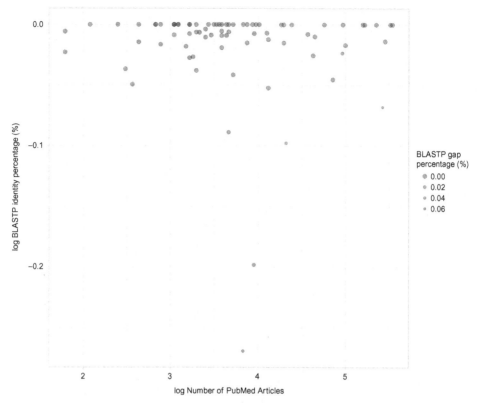

Fig. 2 Number of PubMed articles against the BLASTP identity score of each orthologous RP pairwise alignment. The results for the 79 longest RP protein isoforms are presented in this plot. Both BLASTP identity score and PubMed results were log transformed, and RPS6, which consists a max outlier, was removed for display purposes. The size of points reflects the total size of gaps in RP pairwise alignments. The scatter plot was drawn using ggplot2 R package.

Furthermore, it is worth noting that the level of conservation does not necessarily correlate with the level of literature coverage for RPs (Fig. 2). For example, while coverage for some of the "least-studied" RPs is partly due to less specific name conventions (e.g., UBA52/RPL40 and FAU/RPS30), the highly conserved proteins RPL18a, RPL36a, RPL28, RPS21, RPL38, and RPS28 are covered by less than 15 PubMed entries each, in contrast to the much less conserved RPL29 (76% sequence identity between human and mouse) covered by 46 PubMed entries (Fig. 2).

✦ *Disease involvement of RPs exhibits a complex pattern of associations*

As we have already seen, RP genes and RP haploinsufficiency can lead to ribosomopathies, associated with specific clinical phenotypes and predisposition for cancer development. To better understand the connection of RPs with diseases, we annotated the Titles and Abstracts of all RP-related PubMed articles with the NCBI PubTator software (Wei et al., 2012, 2013b). Specifically, by utilizing PubTator's API (Wei et al., 2016) we have retrieved all Disease-related bioterms from each PubMed article, along with their respective MeSH IDs, as they were determined by the PubTator software. Next, we associated all RPs with MeSH IDs and PubMed results into a complex network, using BioLayout (Goldovsky et al., 2005). The network consists of 82 nodes for RPs and 825 nodes for disease phenotypes (Fig. 3), indicating a very complex association landscape. As expected, disease phenotypes associated with cancer and ribosomopathies are predominant with many RP connections. However, there is also a significant number of unrelated disease phenotypes, for which the involvement of RP functionality is less clear (Fig. 3).

As an example, a more detailed display of connections between RPs and disease phenotypes can be seen in the subnetwork of RPL35a and RPS24 (Fig. 4). Mutations located in the genes of RPS24 and RPL35a are the cause for DBA-3 (OMIM 610629) and DBA-5 (OMIM 612528), respectively. Moreover, in an extensive screening of families with DBA, mutations in RPS24 and RPL35a constituted ~5.5% of all cases (Gazda et al., 2012).

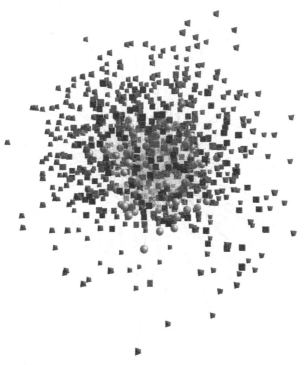

Fig. 3 Network of 82 RPs (*sphere nodes*) and 825 Disease phenotypes (MeSH IDs) (*tetrahydron/cube nodes*). The width of the edges increases based on the number of RP PubMed articles related to the MeSH ID. RPs are represented as sphere-shaped nodes. Tetrahedron-shaped nodes represent unique MeSH ID nodes, which are connected to only one RP node. Cube-shaped nodes represent shared MeSH ID nodes, which are connected to two or more RP nodes. The network was created with BioLayout (Goldovsky et al., 2005).

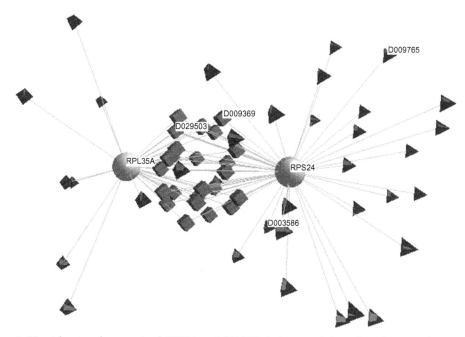

Fig. 4 RPs subnetwork comprised RPS24 and RPL35A (*sphere nodes*), and 59 Disease phenotypes (MeSH IDs) (*tetrahydron/cube nodes*). The width of the edges increases based on the Number of RP PubMed articles related to the MeSH ID. RPS24 and RPL35A are represented as Sphere-shaped nodes. Tetrahydron-shaped nodes represent unique MeSH ID nodes, which are connected with only one RP node. Cube-shaped nodes represent shared MeSH ID nodes, which are connected to both RPS24 and RPL35A nodes. The network was created with BioLayout (Goldovsky et al., 2005).

Unsurprisingly, the MeSH Term "Diamond-Blackfan Anemia" (MeSH ID: D029503) was the most frequent disease-related bioterm in PubMed articles for both RPS24 (18 articles) and RPL35A (14 articles). The second most frequent MeSH Term for both RPS24 (6 articles) and RPL35A (6 articles) was "Neoplasms" (MeSH ID: D009369). In fact, most disease-related bioterms, found in RPS24- and RPL35A-related PubMed articles, are closely related to DBA symptoms and cancer (Fig. 4 and Table 1). Nevertheless, examination of the unique MeSH terms for RPS24 and RPL35A suggests further specialized roles for these RPs. Specifically, RPS24 was found to be upregulated in human endothelial cell line ECV304, following infection with Human Cytomegalovirus (HCMV) (Zhang et al., 2013). Also, RPS24 was downregulated in the hypothalamus of mice with Dicer-knockout POMC-expressing cells (Pro-opiomelanocortin, POMC). These mice are characterized by disrupted metabolism and obesity (Schneeberger et al., 2012). These less studied connections of RPS24 with HCMV infection and obesity were identified through the respective, unique MeSH IDs D003586 and D009765 (Fig. 4 and Table 1).

Table 1 MeSH IDs and terms annotated from RPS24 and/or RPL35A-related PubMed articles (titles and abstracts)

MeSH ID	MeSH term	Ribosomal proteins
D000013	Congenital abnormalities	RPS24 & RPL35A
D000014	Abnormalities, drug-induced	RPS24 & RPL35A
D000740	Anemia	RPS24 & RPL35A
D000748	Anemia, macrocytic	RPS24 & RPL35A
D001855	Bone marrow diseases	RPS24 & RPL35A
D002972	Cleft palate	RPS24 & RPL35A
D006130	Growth disorders	RPS24 & RPL35A
D006330	Heart defects, congenital	RPS24 & RPL35A
D006331	Heart diseases	RPS24 & RPL35A
D006402	Hematologic diseases	RPS24 & RPL35A
D007021	Hypospadias	RPS24 & RPL35A
D007674	Kidney diseases	RPS24 & RPL35A
D009369	Neoplasms	RPS24 & RPL35A
D011488	Protein deficiency	RPS24 & RPL35A
D012010	Red-cell aplasia, pure	RPS24 & RPL35A
D012174	Retinitis pigmentosa	RPS24 & RPL35A
D012734	Disorders of sex development	RPS24 & RPL35A
D029503	Anemia, Diamond-Blackfan	RPS24 & RPL35A
D035583	Rare diseases	RPS24 & RPL35A
C536482	Hereditary renal agenesis	RPS24 & RPL35A
C562798	Transcobalamin I deficiency	RPS24 & RPL35A
C535662	Acromicric dysplasia	RPS24 & RPL35A
C535323	Chromosome 5q deletion syndrome	RPS24 & RPL35A
C536572	Bone marrow failure syndromes	RPS24 & RPL35A
D000312	Adrenal hyperplasia, congenital	RPS24
D001816	Bloom syndrome	RPS24
D002292	Carcinoma, renal cell	RPL35A
D002311	Cardiomyopathy, dilated	RPS24
D002659	Child development disorders, pervasive	RPS24
D003586	Cytomegalovirus infections	RPS24
D004194	Disease	RPS24
D004915	Leukemia, erythroblastic, acute	RPL35A
D006528	Carcinoma, hepatocellular	RPS24
D006963	Hyperphagia	RPS24
D006973	Hypertension	RPS24
D007246	Infertility	RPS24
D008113	Liver neoplasms	RPS24
D009103	Multiple sclerosis	RPL35A
D009140	Musculoskeletal diseases	RPS24
D009190	Myelodysplastic syndromes	RPS24
D009896	Optic atrophy	RPL35A
D010900	Pituitary diseases	RPS24

Table 1 MeSH IDs and terms annotated from RPS24 and/or RPL35A-related PubMed articles (titles and abstracts)—cont'd

MeSH ID	MeSH term	Ribosomal proteins
D012509	Sarcoma	RPS24
D012516	Osteosarcoma	RPL35A
D012769	Shock	RPS24
D016183	Murine-acquired immunodeficiency syndrome	RPL35A
D019337	Hematologic neoplasms	RPS24
D019636	Neurodegenerative diseases	RPS24
D019871	Dyskeratosis congenita	RPS24
D020964	Embryo loss	RPS24
D030342	Genetic diseases, inborn	RPS24
D044882	Glucose metabolism disorders	RPS24
D064420	Drug-related side effects and adverse reactions	RPS24
C537860	Occipital horn syndrome	RPL35A
C537751	Oncogenic osteomalacia	RPL35A
D009765	Obesity	RPS24
C536227	Cyclic neutropenia	RPS24
D053627	Asthenozoospermia	RPS24
C537330	Shwachman syndrome	RPS24

5. CONCLUSIONS AND FUTURE DIRECTIONS

RPs are highly conserved proteins, characterized by newly discovered roles in translational control and cellular homeostasis. Their importance is further highlighted by the delicate control required for the regulation of their gene expression, since perturbations lead to pathological phenotypes. Unexpectedly and despite decades of research in ribosomal structure and function, the level of literature coverage for RPs may vary significantly. Discoveries such as the association of RPS6 phosphorylation with pathological conditions, or the *RPS19* mutations observed in DBA, should in principle explain the reasons why some RPs are better characterized than others. Further research is required to assess this variability, which can also arise from imperfect queries, inconsistent identifiers and disparate funding priorities. The emergence of next-generation sequencing techniques and the continuous growth of relevant data resources are expected to further our understanding of the complex role of RPs in physiological conditions and pathological disorders. Ultimately, tissue-specific patterns of RP gene expression and subcellular stoichiometry will delineate their fundamental association with disease and other phenotypic manifestations.

REFERENCES

Altschul, S.F., Gish, W., Miller, W., Myers, E.W., Lipman, D.J., 1990. Basic local alignment search tool. J. Mol. Biol. 215, 403–410.

Amanatiadou, E.P., Papadopoulos, G.L., Strouboulis, J., Vizirianakis, I.S., 2015. GATA1 and PU.1 bind to ribosomal protein genes in erythroid cells: implications for ribosomopathies. PLoS One 10, e0140077.

Amsterdam, A., Sadler, K.C., Lai, K., Farrington, S., Bronson, R.T., Lees, J.A., Hopkins, N., 2004. Many ribosomal protein genes are cancer genes in zebrafish. PLoS Biol. 2, E139.

Ban, N., Beckmann, R., Cate, J.H., Dinman, J.D., Dragon, F., Ellis, S.R., Lafontaine, D.L., Lindahl, L., Liljas, A., Lipton, J.M., McAlear, M.A., Moore, P.B., Noller, H.F., Ortega, J., Panse, V.G., Ramakrishnan, V., Spahn, C.M., Steitz, T.A., Tchorzewski, M., Tollervey, D., Warren, A.J., Williamson, J.R., Wilson, D., Yonath, A., Yusupov, M., 2014. A new system for naming ribosomal proteins. Curr. Opin. Struct. Biol. 24, 165–169.

Barakat, A., Szick-Miranda, K., Chang, I.-F., Guyot, R., Blanc, G., Cooke, R., Delseny, M., Bailey-Serres, J., 2001. The organization of cytoplasmic ribosomal protein genes in the arabidopsis genome. Plant Physiol. 127, 398–415.

Bellodi, C., Kopmar, N., Ruggero, D., 2010. Deregulation of oncogene-induced senescence and p53 translational control in X-linked dyskeratosis congenita. EMBO J. 29, 1865–1876.

Boria, I., Garelli, E., Gazda, H.T., Aspesi, A., Quarello, P., Pavesi, E., Ferrante, D., Meerpohl, J.J., Kartal, M., Da Costa, L., Proust, A., Leblanc, T., Simansour, M., Dahl, N., Fröjmark, A.-S., Pospisilova, D., Cmejla, R., Beggs, A.H., Sheen, M.R., Landowski, M., Buros, C.M., Clinton, C.M., Dobson, L.J., Vlachos, A., Atsidaftos, E., Lipton, J.M., Ellis, S.R., Ramenghi, U., Dianzani, I., 2010. The ribosomal basis of diamond-blackfan anemia: mutation and database update. Hum. Mutat. 31, 1269–1279.

Carroll, A.J., Heazlewood, J.L., Ito, J., Millar, A.H., 2008. Analysis of the Arabidopsis cytosolic ribosome proteome provides detailed insights into its components and their post-translational modification. Mol. Cell. Proteomics 7, 347–369.

Dahl, N., Draptchinskaia, N., Gustavsson, P., Andersson, B., Pettersson, M., Willig, T.-N., Dianzani, I., Ball, S., Tchernia, G., Klar, J., Matsson, H., Tentler, D., Mohandas, N., Carlsson, B., 1999. The gene encoding ribosomal protein S19 is mutated in Diamond-Blackfan anaemia. Nat. Genet. 21, 169–175.

Danilova, N., Sakamoto, K.M., Lin, S., 2008. Ribosomal protein S19 deficiency in zebrafish leads to developmental abnormalities and defective erythropoiesis through activation of p53 protein family. Blood 112, 5228–5237.

De Keersmaecker, K., Atak, Z.K., Li, N., Vicente, C., Patchett, S., Girardi, T., Gianfelici, V., Geerdens, E., Clappier, E., Porcu, M., Lahortiga, I., Lucà, R., Yan, J., Hulselmans, G., Vranckx, H., Vandepoel, R., Sweron, B., Jacobs, K., Mentens, N., Wlodarska, I., Cauwelier, B., Cloos, J., Soulier, J., Uyttebroeck, A., Bagni, C., Hassan, B.A., Vandenberghe, P., Johnson, A.W., Aerts, S., Cools, J., 2013. Exome sequencing identifies mutation in CNOT3 and ribosomal genes RPL5 and RPL10 in T-cell acute lymphoblastic leukemia. Nat. Genet. 45, 186–190.

De Keersmaecker, K., Sulima, S.O., Dinman, J.D., 2015. Ribosomopathies and the paradox of cellular hypo- to hyperproliferation. Blood 125, 1377–1382.

Dez, C., Tollervey, D., 2004. Ribosome synthesis meets the cell cycle. Curr. Opin. Microbiol. 7, 631–637.

Dutt, S., Narla, A., Lin, K., Mullally, A., Abayasekara, N., Megerdichian, C., Wilson, F.H., Currie, T., Khanna-Gupta, A., Berliner, N., Kutok, J.L., Ebert, B.L., 2011. Haploinsufficiency for ribosomal protein genes causes selective activation of p53 in human erythroid progenitor cells. Blood 117, 2567–2576.

Ebert, B.L., Pretz, J., Bosco, J., Chang, C.Y., Tamayo, P., Galili, N., Raza, A., Root, D.E., Attar, E., Ellis, S.R., Golub, T.R., 2008. Identification of RPS14 as a 5q-syndrome gene by RNA interference screen. Nature 451, 335–339.

Ellis, S.R., Lipton, J.M., 2008. Chapter 8 Diamond Blackfan anemia: a disorder of red blood cell development. Curr. Top. Dev. Biol. 82, 217–241.

FANTOM Consortium and the RIKEN PMI and CLST (DGT), Forrest, A.R., Kawaji, H., Rehli, M., Baillie, J.K., de Hoon, M.J., Haberle, V., Lassmann, T., Kulakovskiy, I.V., Lizio, M., Itoh, M., Andersson, R., Mungall, C.J., Meehan, T.F., Schmeier, S., Bertin, N., Jørgensen, M., Dimont, E., Arner, E., Schmidl, C., Schaefer, U., Medvedeva, Y.A., Plessy, C., Vitezic, M., Severin, J., Semple, C.A., Ishizu, Y., Young, R.S., Francescatto, M., Alam, I., Albanese, D., Altschuler, G.M., Arakawa, T., Archer, J.A., Arner, P., Babina, M., Rennie, S., Balwierz, P.J., Beckhouse, A.G., Pradhan-Bhatt, S., Blake, J.A., Blumenthal, A., Bodega, B., Bonetti, A., Briggs, J., Brombacher, F., Burroughs, A.M., Califano, A., Cannistraci, C.V., Carbajo, D., Chen, Y., Chierici, M., Ciani, Y., Clevers, H.C., Dalla, E., Davis, C.A., Detmar, M., Diehl, A.D., Dohi, T., Drabløs, F., Edge, A.S.B., Edinger, M., Ekwall, K., Endoh, M., Enomoto, H., Fagiolini, M., Fairbairn, L., Fang, H., Farach-Carson, M.C., Faulkner, G.J., Favorov, A.V., Fisher, M.E., Frith, M.C., Fujita, R., Fukuda, S.,

Furlanello, C., Furino, M., Furusawa, J., Geijtenbeek, T.B., Gibson, A.P., Gingeras, T., Goldowitz, D., Gough, J., Guhl, S., Guler, R., Gustincich, S., Ha, T.J., Hamaguchi, M., Hara, M., Harbers, M., Harshbarger, J., Hasegawa, A., Hasegawa, Y., Hashimoto, T., Herlyn, M., Hitchens, K.J., Ho Sui, S.J., Hofmann, O.M., Hoof, I., Hori, F., Huminiecki, L., Iida, K., Ikawa, T., Jankovic, B.R., Jia, H., Joshi, A., Jurman, G., Kaczkowski, B., Kai, C., Kaida, K., Kaiho, A., Kajiyama, K., Kanamori-Katayama, M., Kasianov, A.S., Kasukawa, T., Katayama, S., Kato, S., Kawaguchi, S., Kawamoto, H., Kawamura, Y.I., Kawashima, T., Kempfle, J.S., Kenna, T.J., Kere, J., Khachigian, L.M., Kitamura, T., Klinken, S.P., Knox, A.J., Kojima, M., Kojima, S., Kondo, N., Koseki, H., Koyasu, S., Krampitz, S., Kubosaki, A., Kwon, A.T., Laros, J.F.J., Lee, W., Lennartsson, A., Li, K., Lilje, B., Lipovich, L., Mackay-Sim, A., Manabe, R., Mar, J.C., Marchand, B., Mathelier, A., Mejhert, N., Meynert, A., Mizuno, Y., de Lima Morais, D.A., Morikawa, H., Morimoto, M., Moro, K., Motakis, E., Motohashi, H., Mummery, C.L., Murata, M., Nagao-Sato, S., Nakachi, Y., Nakahara, F., Nakamura, T., Nakamura, Y., Nakazato, K., van Nimwegen, E., Ninomiya, N., Nishiyori, H., Noma, S., Noma, S., Noazaki, T., Ogishima, S., Ohkura, N., Ohimiya, H., Ohno, H., Ohshima, M., Okada-Hatakeyama, M., Okazaki, Y., Orlando, V., Ovchinnikov, D.A., Pain, A., Passier, R., Patrikakis, M., Persson, H., Piazza, S., Prendergast, J.G., Rackham, O.J., Ramilowski, J.A., Rashid, M., Ravasi, T., Rizzu, P., Roncador, M., Roy, S., Rye, M.B., Saijyo, E., Sajantila, A., Saka, A., Sakaguchi, S., Sakai, M., Sato, H., Savvi, S., Saxena, A., Schneider, C., Schultes, E.A., Schulze-Tanzil, G.G., Schwegmann, A., Sengstag, T., Sheng, G., Shimoji, H., Shimoni, Y., Shin, J.W., Simon, C., Sugiyama, D., Sugiyama, T., Suzuki, M., Suzuki, N., Swoboda, R.K., 't Hoen, P.A., Tagami, M., Takahashi, N., Takai, J., Tanaka, H., Tatsukawa, H., Tatum, Z., Thompson, M., Toyodo, H., Toyoda, T., Valen, E., van de Wetering, M., van den Berg, L.M., Verado, R., Vijayan, D., Vorontsov, I.E., Wasserman, W.W., Watanabe, S., Wells, C.A., Winteringham, L.N., Wolvetang, E., Wood, E.J., Yamaguchi, Y., Yamamoto, M., Yoneda, M., Yonekura, Y., Yoshida, S., Zabierowski, S.E., Zhang, P.G., Zhao, X., Zucchelli, S., Summers, K.M., Suzuki, H., Daub, C.O., Kawai, J., Heutink, P., Hide, W., Freeman, T.C., Lenhard, B., Bajic, V.B., Taylor, M.S., Makeev, V.J., Sandelin, A., Hume, D.A., Carninci, P., Hayashizaki, Y., 2014. A promoter-level mammalian expression atlas. Nature 507, 462–470.

Gazda, H.T., Sheen, M.R., Vlachos, A., Choesmel, V., O'Donohue, M.-F., Schneider, H., Darras, N., Hasman, C., Sieff, C.A., Newburger, P.E., Ball, S.E., Niewiadomska, E., Matysiak, M., Zaucha, J.M., Glader, B., Niemeyer, C., Meerpohl, J.J., Atsidaftos, E., Lipton, J.M., Gleizes, P.-E., Beggs, A.H., 2008. Ribosomal protein L5 and L11 mutations are associated with cleft palate and abnormal thumbs in Diamond-Blackfan anemia patients. Am. J. Hum. Genet. 83, 769–780.

Gazda, H.T., Preti, M., Sheen, M.R., O'Donohue, M.-F., Vlachos, A., Davies, S.M., Kattamis, A., Doherty, L., Landowski, M., Buros, C., Ghazvinian, R., Sieff, C.A., Newburger, P.E., Niewiadomska, E., Matysiak, M., Glader, B., Atsidaftos, E., Lipton, J.M., Gleizes, P.-E., Beggs, A.H., 2012. Frameshift mutation in p53 regulator *RPL26* is associated with multiple physical abnormalities and a specific pre-ribosomal RNA processing defect in diamond-blackfan anemia. Hum. Mutat. 33, 1037–1044.

Goldovsky, L., Cases, I., Enright, A.J., Ouzounis, C.A., 2005. BioLayout(Java): versatile network visualisation of structural and functional relationships. Appl. Bioinforma. 4, 71–74.

Golomb, L., Volarevic, S., Oren, M., 2014. p53 and ribosome biogenesis stress: the essentials. FEBS Lett. 588, 2571–2579.

Guimaraes, J.C., Zavolan, M., 2016. Patterns of ribosomal protein expression specify normal and malignant human cells. Genome Biol. 17, 236.

Horos, R., IJspeert, H., Pospisilova, D., Sendtner, R., Andrieu-Soler, C., Taskesen, E., Nieradka, A., Cmejla, R., Sendtner, M., Touw, I.P., von Lindern, M., 2012. Ribosomal deficiencies in Diamond-Blackfan anemia impair translation of transcripts essential for differentiation of murine and human erythroblasts. Blood 119, 262–272.

Kandoth, C., McLellan, M.D., Vandin, F., Ye, K., Niu, B., Lu, C., Xie, M., Zhang, Q., McMichael, J.F., Wyczalkowski, M.A., Leiserson, M.D.M., Miller, C.A., Welch, J.S., Walter, M.J., Wendl, M.C., Ley, T.J., Wilson, R.K., Raphael, B.J., Ding, L., 2013. Mutational landscape and significance across 12 major cancer types. Nature 502, 333–339.

Kellis, M., Birren, B.W., Lander, E.S., 2004. Proof and evolutionary analysis of ancient genome duplication in the yeast Saccharomyces cerevisiae. Nature 428, 617–624.

Kim, J.-H., You, K.-R., Kim, I.H., Cho, B.-H., Kim, C.-Y., Kim, D.-G., 2004. Over-expression of the ribosomal protein L36a gene is associated with cellular proliferation in hepatocellular carcinoma. Hepatology 39, 129–138.

Kondrashov, N., Pusic, A., Stumpf, C.R., Shimizu, K., Hsieh, A.C., Xue, S., Ishijima, J., Shiroishi, T., Barna, M., 2011. Ribosome-mediated specificity in Hox mRNA translation and vertebrate tissue patterning. Cell 145, 383–397.

Kressler, D., Hurt, E., Bassler, J., 2010. Driving ribosome assembly. Biochim. Biophys. Acta 1803, 673–683.

Krieg, J., Hofsteenge, J., Thomas, G., 1988. Identification of the 40 S ribosomal protein S6 phosphorylation sites induced by cycloheximide. J. Biol. Chem. 263, 11473–11477.

Lafontaine, D.L.J., 2015. Noncoding RNAs in eukaryotic ribosome biogenesis and function. Nat. Struct. Mol. Biol. 22, 11–19.

Lee, S.-W., Berger, S.J., Martinović, S., Pasa-Tolić, L., Anderson, G.A., Shen, Y., Zhao, R., Smith, R.D., 2002. Direct mass spectrometric analysis of intact proteins of the yeast large ribosomal subunit using capillary LC/FTICR. Proc. Natl. Acad. Sci. USA 99, 5942–5947.

Ludwig, L.S., Gazda, H.T., Eng, J.C., Eichhorn, S.W., Thiru, P., Ghazvinian, R., George, T.I., Gotlib, J.R., Beggs, A.H., Sieff, C.A., Lodish, H.F., Lander, E.S., Sankaran, V.G., 2014. Altered translation of GATA1 in Diamond-Blackfan anemia. Nat. Med. 20, 748–753.

Marygold, S.J., Roote, J., Reuter, G., Lambertsson, A., Ashburner, M., Millburn, G.H., Harrison, P.M., Yu, Z., Kenmochi, N., Kaufman, T.C., Leevers, S.J., Cook, K.R., 2007. The ribosomal protein genes and Minute loci of Drosophila melanogaster. Genome Biol. 8, R216.

McGowan, K.A., Li, J.Z., Park, C.Y., Beaudry, V., Tabor, H.K., Sabnis, A.J., Zhang, W., Fuchs, H., de Angelis, M.H., Myers, R.M., Attardi, L.D., Barsh, G.S., 2008. Ribosomal mutations cause p53-mediated dark skin and pleiotropic effects. Nat. Genet. 40, 963–970.

Meyuhas, O., 2015. Ribosomal protein S6 phosphorylation. Int. Rev. Cell Mol. Biol. 320, 41–73. https://doi.org/10.1016/bs.ircmb.2015.07.006.

Nakhoul, H., Ke, J., Zhou, X., Liao, W., Zeng, S.X., Lu, H., 2014. Ribosomopathies: mechanisms of disease. Clin. Med. Insights Blood Disord. 7, 7–16.

Nicolas, E., Parisot, P., Pinto-Monteiro, C., de Walque, R., De Vleeschouwer, C., Lafontaine, D.L.J., 2016. Involvement of human ribosomal proteins in nucleolar structure and p53-dependent nucleolar stress. Nat. Commun. 7, 11390.

Odintsova, T.I., Müller, E.-C., Ivanov, A.V., Egorov, T.A., Bienert, R., Vladimirov, S.N., Kostka, S., Otto, A., Wittmann-Liebold, B., Karpova, G.G., 2003. Characterization and analysis of posttranslational modifications of the human large cytoplasmic ribosomal subunit proteins by mass spectrometry and Edman sequencing. J. Protein Chem. 22, 249–258.

Plank, T.-D.M., Kieft, J.S., 2012. The structures of nonprotein-coding RNAs that drive internal ribosome entry site function. Wiley Interdiscip. Rev. RNA 3, 195–212.

Rao, S., Lee, S.-Y., Gutierrez, A., Perrigoue, J., Thapa, R.J., Tu, Z., Jeffers, J.R., Rhodes, M., Anderson, S., Oravecz, T., Hunger, S.P., Timakhov, R.A., Zhang, R., Balachandran, S., Zambetti, G.P., Testa, J.R., Look, A.T., Wiest, D.L., 2012. Inactivation of ribosomal protein L22 promotes transformation by induction of the stemness factor, Lin28B. Blood 120, 3764–3773.

Ray, P.S., Grover, R., Das, S., 2006. Two internal ribosome entry sites mediate the translation of p53 isoforms. EMBO Rep. 7, 404–410.

Ruggero, D., Shimamura, A., 2014. Marrow failure: a window into ribosome biology. Blood 124, 2784–2792.

Sankaran, V.G., Ghazvinian, R., Do, R., Thiru, P., Vergilio, J.-A., Beggs, A.H., Sieff, C.A., Orkin, S.H., Nathan, D.G., Lander, E.S., Gazda, H.T., 2012. Exome sequencing identifies GATA1 mutations resulting in Diamond-Blackfan anemia. J. Clin. Invest. 122, 2439–2443.

Schneeberger, M., Altirriba, J., Garcia, A., Esteban, Y., Castaño, C., García-Lavandeira, M., Alvarez, C.V., Gomis, R., Claret, M., 2012. Deletion of miRNA processing enzyme Dicer in POMC-expressing cells leads to pituitary dysfunction, neurodegeneration and development of obesity. Mol. Metab. 12, 74–85.

Slavov, N., Semrau, S., Airoldi, E., Budnik, B., van Oudenaarden, A., 2015. Differential stoichiometry among core ribosomal proteins. Cell Rep. 13, 865–873.

Sonenberg, N., Hinnebusch, A.G., 2009. Regulation of translation initiation in eukaryotes: mechanisms and biological targets. Cell 136, 731–745.

Spriggs, K.A., Stoneley, M., Bushell, M., Willis, A.E., 2008. Re-programming of translation following cell stress allows IRES-mediated translation to predominate. Biol. Cell. 100, 27–38.

Stoneley, M., Paulin, F.E., Le Quesne, J.P., Chappell, S.A., Willis, A.E., 1998. C-Myc 5' untranslated region contains an internal ribosome entry segment. Oncogene 16, 423–428.

Sulima, S.O., Patchett, S., Advani, V.M., De Keersmaecker, K., Johnson, A.W., Dinman, J.D., 2014. Bypass of the pre-60S ribosomal quality control as a pathway to oncogenesis. Proc. Natl. Acad. Sci. USA 111, 5640–5645.

Viswanathan, S.R., Powers, J.T., Einhorn, W., Hoshida, Y., Ng, T.L., Toffanin, S., O'Sullivan, M., Lu, J., Phillips, L.A., Lockhart, V.L., Shah, S.P., Tanwar, P.S., Mermel, C.H., Beroukhim, R., Azam, M., Teixeira, J., Meyerson, M., Hughes, T.P., Llovet, J.M., Radich, J., Mulligan, C.G., Golub, T.R., Sorensen, P.H., Daley, G.Q., 2009. Lin28 promotes transformation and is associated with advanced human malignancies. Nat. Genet. 41, 843–848.

Vlachos, A., Rosenberg, P.S., Atsidaftos, E., Alter, B.P., Lipton, J.M., 2012. Incidence of neoplasia in Diamond Blackfan anemia: a report from the Diamond Blackfan Anemia Registry. Blood 119, 3815–3819.

Wang, M., Hu, Y., Stearns, M.E., 2009. RPS2: a novel therapeutic target in prostate cancer. J. Exp. Clin. Cancer Res. 28, 6.

Wei, C.-H., Harris, B.R., Li, D., Berardini, T.Z., Huala, E., Kao, H.-Y., Lu, Z., 2012. Accelerating literature curation with text-mining tools: a case study of using PubTator to curate genes in PubMed abstracts. Database 2012, bas041-bas041.

Wei, S., Chen, X., McGraw, K., Zhang, L., Komrokji, R., Clark, J., Caceres, G., Billingsley, D., Sokol, L., Lancet, J., Fortenbery, N., Zhou, J., Eksioglu, E.A., Sallman, D., Wang, H., Epling-Burnette, P.K., Djeu, J., Sekeres, M., Maciejewski, J.P., List, A., 2013a. Lenalidomide promotes p53 degradation by inhibiting MDM2 auto-ubiquitination in myelodysplastic syndrome with chromosome 5q deletion. Oncogene 32, 1110–1120.

Wei, C.-H., Kao, H.-Y., Lu, Z., 2013b. PubTator: a web-based text mining tool for assisting biocuration. Nucleic Acids Res. 41, W518–W522.

Wei, C.-H., Leaman, R., Lu, Z., 2016. Beyond accuracy: creating interoperable and scalable text-mining web services. Bioinformatics 32, 1907–1910.

Wong, Q.W.-L., Li, J., Ng, S.R., Lim, S.G., Yang, H., Vardy, L.A., 2014. RPL39L is an example of a recently evolved ribosomal protein paralog that shows highly specific tissue expression patterns and is upregulated in ESCs and HCC tumors. RNA Biol. 11, 33–41.

Xue, S., Barna, M., 2012. Specialized ribosomes: a new frontier in gene regulation and organismal biology. Nat. Rev. Mol. Cell Biol. 13, 355–369.

Xue, S., Barna, M., 2015. Cis-regulatory RNA elements that regulate specialized ribosome activity. RNA Biol. 12, 1083–1087.

Xue, S., Tian, S., Fujii, K., Kladwang, W., Das, R., Barna, M., 2015. RNA regulons in Hox 5' UTRs confer ribosome specificity to gene regulation. Nature 517, 33–38.

Yelick, P.C., Trainor, P.A., 2015. Ribosomopathies: global process, tissue specific defects. Rare Dis. (Austin, Tex.) 3, e1025185.

Yu, Y., Ji, H., Doudna, J.A., Leary, J.A., 2009. Mass spectrometric analysis of the human 40S ribosomal subunit: native and HCV IRES-bound complexes. Protein Sci. 14, 1438–1446.

Zeidan, Q., Wang, Z., De Maio, A., Hart, G.W., 2010. O-GlcNAc cycling enzymes associate with the translational machinery and modify core ribosomal proteins. Mol. Biol. Cell 21, 1922–1936.

Zhang, Y., Ma, W., Mo, X., Zhao, H., Zheng, H., Ke, C., Zheng, W., Tu, Y., Zhang, Y., 2013. Erratum: differential expressed genes in ECV304 endothelial-like cells infected with human cytomegalovirus. Afr. Health Sci. 13, 864–879.

CHAPTER 4

Big Data, Artificial Intelligence, and Machine Learning in Neurotrauma

Denes V. Agoston
Department of Anatomy, Physiology and Genetics, Uniformed Services University, Bethesda, MD, United States

1. INTRODUCTION

Big Data (BD) stands for extremely large, complex structured, semistructured or unstructured data sets that they cannot be analyzed using traditional data processing applications (Choudhury et al., 2014; Dilsizian and Siegel, 2014; Baro et al., 2015; White and Ford, 2015; Chen et al., 2016; Flechet et al., 2016; Lebo et al., 2016; Pastur-Romay et al., 2016; Peng et al., 2016; Bzdok and Yeo, 2017; Mooney and Pejaver, 2017; Kissin, 2018; Talboom and Huentelman, 2018). BD may be analyzed or "mined" using special computational tools, text mining (TM), artificial intelligence (AI), or Machine Learning (ML) to reveal patterns, trends, and associations. At the emergence of the field in the 1990s, BD was characterized or defined by the "3 Vs"; Volume, Variety, Velocity. As the field has expanded, two additional characteristics, Variability and Veracity, were added, and BD was characterized by 5 Vs in the early 2000s. Recently, additional Vs, Validity, Vulnerability, Volatility, Visualization, and Value, have been added and so BD is now characterized by the 10 Vs. This chapter will first introduce each of the 10 Vs, followed by examples for each of the Vs in TBI. It will then introduce and discuss the concepts of structured vs. unstructured data—textual information, like journal articles or reports, and introduce and discuss the tools—TM, AI, and AI that can analyze and interpret BD and will present examples for potential uses of TM, AI, and ML in TBI. Finally, the current status of the TBI field for BD approaches, the opportunities, and challenges ahead will be discussed.

2. BIG DATA: CHARACTERISTICS, DEFINITIONS, AND EXAMPLES

Volume: Volume is the most important characteristic of BD and it refers to the amount of data (Wang and Krishnan, 2014). The ability to collect all kinds of data and store them has resulted in the deluge of data in all aspects of our lives. There were ~1.2 TRILLION photos (rather call them images) taken in 2017. Assuming that the average size of a digital image is ~1 megabyte, digital images represent a stunning 1200 petabyte of data. Most images are stored on several devices (smart phones, computers, tablets, etc.), and the total number of stored images was 5 TRILLION in 2017. This translates into 6000 petabyte of stored data. Or, in 2017, there were ~300h of video uploaded to YouTube in every MINUTE. Calculating with ~ 1 gigabytes per hour of video (low resolution), this translates into ~5 GB data per minute, 3000 GB per hour, 3000 GB × 276,000 h = 26,280,000 GB which is ~26 petabyte per year. Biomedical data volume is on the same magnitude, and has been growing exponentially due to data rich technologies—the various "omics," genomics, proteomics, metabolomics, etc. (Choudhury et al., 2014; Talboom and Huentelman, 2018; Andreu-Perez et al., 2015). Various forms of imaging generate most of the data, a single, complete high-resolution neuroimaging study is in the terabyte range, two or three orders of magnitude larger than the entire human genome. A typical research article in a scientific journal, with images is in the 1–10 megabyte range and the entire currently published biomedical literature is in the petabyte range.

Variety: Variety means that data are in different formats, structured but mostly unstructured (Baro et al., 2015; Kansagra et al., 2016; Webb-Vargas et al., 2017). Structured data examples are relational databases, numbers, or units arranged in a way that can be easily interrogated for correlations using traditional methods. The majority, more than 70% of all data is text, images, numeric data, graphs, tables, multimedia, video, or the combination of these. This data format is called unstructured, i.e., data are not organized in any predefined manner and cannot be analyzed with traditional approaches. By definition, unstructured data contain a significant amount of uncertain and imprecise data. Scientific and clinical reports, case studies are the prime examples of unstructured data; they contain a combination of raw text, images, videos, numerical data, etc. Extracting, interpreting data, finding correlations, understanding and interpreting a scientific report that contains unstructured data is a significant challenge for BD approaches. In addition to structured and unstructured data, data can be semistructured, i.e., data are not in a traditional relational database, but the data have some organizational properties, e.g., data are in NoSQL databases, which make it easier to analyze the data.

Velocity: Velocity means that most data are in motion, data have temporal dimension (Baro et al., 2015; Sebaa et al., 2018). Accordingly, data need to be placed in the correct temporal context even when they represent a single data point. Even a single datum collected at a preselected experimental (end)point has to be defined in terms of time so to enable analysis, comparison, and correlation with other similarly "static" data points representing different time points. The same applies to data collected using periodic sampling, i.e., using multiple time points, they have to be associated with the time of collection enabling to define their velocity enabling correctly analyzing and correlating them with other data in the same temporal dimension.

Variability: Variability refers to the number of inconsistencies in the data, the multiple data types, data sources and dimensions, and variable speed of data collection and deposition (Rodriguez et al., 2018). Variability is especially important in textual data as the meaning of words can be vastly different in various textual context. Accordingly, specific programs need to be developed that not only directly translates the meaning of a single word, but by analyzing the widest possible textual context it can provide the precise meaning of the word in the actual context it is used. This is how Google has created Google Translate, currently able to translate back and forth between more than 100 different languages.

Veracity: Veracity means the degree to which the data are accurate and precise (Baro et al., 2015). Low veracity means that data are uneven in quality, incomplete, or ambiguous or deceptive. Veracity thus refers to the trustworthiness of the data. Veracity can be defined as the degree to which data are accurate. Examples of factors affecting veracity of data include "biases" (due to subjectivity or poor statistical analysis); "noise" (insufficient separation of significant data from insignificant data); "falsification" (manipulation of primary data).

Validity: Validity is defined as the degree of accuracy and correctness of the data for the intended use (Devinsky et al., 2016; McIntyre et al., 2014; Wasser et al., 2015). "Raw" data are messy, inaccurate, so validity of data needs to be established, typically by "cleansing" the primary data. As data volume grows and analytical tools evolve, this will be likely a lesser issue but consistency in data quality and using common definitions will remain critical.

Vulnerability: Vulnerability of data means that data can be hacked and manipulated (Andreu-Perez et al., 2015; Toga and Dinov, 2015). Bigger the data, more likely that data will be breached especially if there is a commercial or financial interest associated with data.

Volatility: Volatility refers to the length of time data is valid, i.e., how long should it be stored (Wang and Krishnan, 2014; Wasser et al., 2015; Dinov, 2016a; Van Horn and Toga, 2014). Data can be stored indefinitely, but how useful are "old" data collected, e.g., by using outdated methods or instruments, is an important issue. The old means or ways may no any longer be accepted due to various reasons, such as data quality. Importantly, as the stored data volume grows, storage and retrieval as well as analysis will also be more time consuming and complicated.

Visualization: Visualization means to transform data into graphs, plots, etc. that can help to summarize, analyze BD for humans (Bolouri et al., 2016; Dinov, 2016b; Dinov et al., 2016). Visualizing BD is currently challenging due to limitations in processing power, memory technology, and time required. Visualizing BD cannot use traditional graphs designed for few data points. Rather visualization of BD can be attempted using tree maps, cone trees, etc.

Value: Value means the usefulness of data for the intended purpose (Chen et al., 2016; Dinov, 2016a; Cahsai et al., 2015; Vallmuur et al., 2016). Great value can be found in BD but BD alone cannot guarantee that there is a value in the available BD. Determining the—potential—value of the available BD to deliver value—return on investment—is a substantial challenge in using BD.

3. TRAUMATIC BRAIN INJURY (TBI)

TBI has been deservedly called "The most complex disease of the most complex organ" (Wheble and Menon, 2016). Importantly, TBI is a not a single disease entity but rather it consists a whole spectrum of severities and pathological processes leading to both short-term symptoms and long-term consequences (Agoston and Elsayed, 2012; Duhaime et al., 2012; Esselman and Uomoto, 1995; McKee et al., 2013; Siddiqui et al., 2017). Arguably, there is little in common between mild TBI (clinically called concussion) and severe TBI in terms of symptomatology, pathological processes, and certainly outcomes. Suffering severe and even moderate TBI may increase the risk of Alzheimer's Disease by as much as 4 times (Jellinger, 2004; Portbury and Adlard, 2015; Sivanandam and Thakur, 2012). The most frequent—the mild form of TBI, which is frequently repeated can significantly—up to 5-fold—increase the risk for developing chronic trau-matic encephalopathy (CTE) (Hasoon, 2017; Hay et al., 2016; Kiernan et al., 2015; Lucke-Wold et al., 2014).

The normal, healthy human brain is hugely complex, it contains ~100 billion neu-rons and 10 times more glial cells and billions and billions of other cell types (Fundamental Neuroscience, 2013). Endothelial cells that line cerebral vasculature, and cells that form structures like the choroid plexus, trabeculae, dura, and pia mater are numbered in the billions. Each neuron uses a combination of classical and peptide neurotransmitters, forms ~7000 synapses, and every synapse has an array of different ion channels and receptors in various configurations. Critically, neurons and other cell types, cells as well as the neuronal processes, dendrites, and axons are arranged in ultraprecise, highly conserved three-dimensional space, adding another critical layer of complexity. Moreover, these structures are not frozen in time, synapses are eliminated or enhanced, the expression of ion channels and receptors changes over time. The molecular complexity of the brain is similarly stunning; more than 80% of all genes are expressed in the CNS and ~20% of the genome is brain-specific. The human brain proteome, containing myriads of cell surface and signaling molecules, receptors, transporters, etc., is the most elaborate proteome of any organ system. This superbly complex, intricate, and structurally and functionally extremely well-balanced system is suddenly unhinged by the impact of physical forces causing structural damage (primary injury process). This ultrashort phase of TBI is followed by an extremely complex, dynamically changing/time dependent and lasting biological response, the secondary injury process. While the primary injury can be only prevented by avoiding TBI or mitigated by protective gear, e.g. helmets, once happened it is the secondary injury process, e.g., metabolic crisis, inflammation, etc. that needs to be addressed. Further variability—data—is that the biological response changes over time, i.e., has temporal profile. Trying to calculate all of the potential combinations between physical forces, primary and secondary injury processes, structures and molecules affected and/or involved, and their temporal differences could easily produce an enormous data volume in the order of magnitude of one googol (10100).

There appears to be correlation between physical forces, severity of injury, and the secondary injury process (Agoston and Elsayed, 2012; Agoston and Kamnaksh, 2015; Agoston and Langford, 2017). Mild TBI or concussion predominantly causes transient metabolic changes (Bergsneider et al., 2000, 2001; Moore et al., 2000; Vespa et al., 2005), whereas severe acceleration/deceleration in TBI results in substantial structural damage, vascular (Bell et al., 2010; Golding, 2002; Logsdon et al., 2015) and axonal (Buki and Povlishock, 2006; Hill et al., 2016; Johnson et al., 2013; Povlishock, 1992, 1993; Povlishock and Christman, 1995) injuries followed by complex secondary injury processes including inflammation (Finnie, 2013; Kumar and Loane, 2012; Simon et al., 2017). However, the exact correlation between physical forces and the biological response is not known and it is likely far more complex than the simple value of g-forces. Directionality of forces plays a hugely important role in injury severity, but we have very little data about this critical factor (Johnson et al., 2013; Browne et al., 2011).

4. BIG DATA IN TBI

BD approaches in biomedical research, including TBI, are still in their infancy for several reasons. One is because the most important tool—text mining—that can extract high-quality information from, e.g., publications, medical records, and reports is still not advanced enough. The other reasons include varying data standards, and that critical data remain unreported ("dark data," see later).

Volume: There are ∼40,000 publications TBI-related papers in peer-reviewed journals as of early 2018 (legacy data). All of them are text heavy and in unstructured data format. A typical publication—depending on the content—is in the megabyte range, so the entire TBI literature is in the terabyte range. However, this is without full-size, high-resolution images, but TBI literature frequently has large number of images and along with the supplemental information publications can reach gigabyte sizes. The entire current biomedical literature is in the several petabyte range. In addition to published scientific papers and reports, called legacy data, clinical and some experimental TBI research collects very large datasets, mostly in unstructured format (Bigler, 2001; Irimia et al., 2012; Mu et al., 2016; Shulman and Strashun, 2009). An example of such data collection is neurocritical care monitoring (Agoston, 2015, 2017). Along with routine or near-routine monitoring, data derived from EEGs, imaging, etc. results in huge amounts of data collected from each patient.

Variety: Legacy TBI data—published papers—are all in unstructured, mostly text heavy format. Fresh data collected, e.g., in Neurointensive Care Units (NICU) during patient monitoring is a good example of both volume and variety. The—streaming—or real-time data include vital signs (heart, pulse, breath rates) physiological and biochemical parameters such as cerebral perfusion pressure, cerebral blood flow, brain tissue oxygenation, intracranial pressure, changes in intracranial glucose metabolism, etc. that are semistructured (Agoston, 2017). The addition of notes, reports that are text heavy, to the semistructured data illustrates very well the variety of data in TBI (Ferguson et al., 2014; Wallis et al., 2013).

Velocity: Just as in any other field, data collected in TBI research are in motion, all data have a temporal dimension (Bouzat et al., 2014; Matz and Pitts, 1997; Shen et al., 2007). Example of data velocity is again data collected at the NICU with real-time streaming data, e.g., physiological/vital information, heart and breath rates, intracranial pressure, blood pressure, etc. The importance of velocity having the temporal pattern of data is critical; it can provide the temporal pattern of changes, "trends," i.e., improving or worsening clinical parameters over time. As the costs of collecting and storing data are getting less expensive, near real-time or real-time data streaming is becoming increasingly common. Velocity exists even when data are collected from experimental studies with seemingly static time points. Even a single datum collected at a preselected experimental (end)point has to be defined in terms of time so to enable analysis, comparison, and correlation with other similarly "static" data points representing different time points. The same applies to data collected using periodic sampling, i.e., using multiple time points, they have to be associated with the time of collection enabling to define their velocity enabling correctly analyzing and correlating them with other data.

Variability: Examples of variability in the TBI field include, e.g., a parameter (e.g., biochemical measurement, or 3D location of a structure in neuroimaging) that is expected to be the same varies between locations due to variability caused by humans or machines (Andriessen et al., 2011; Eierud et al., 2014; Manley et al., 2010; Manor et al., 2008; Smith, 2008). Other examples of variability include when researchers or physicians use different machines for measuring the same parameter or human variability, using the same machines at different locations. Variability can have significant impact on the overall quality of data and on our ability to homogenize data, i.e., not knowing how representative each data points is. Variability is a critical issue in biomedical including TBI research; it is responsible for the notoriously low level of reproducibility of biomedical studies (Addona et al., 2009; De Guio et al., 2016; Terry and Desiderio, 2003).

Veracity: Data are worthless if they are not accurate and the result can be life threatening, e.g., in clinical TBI if the decision is automated, e.g., based on unsupervised machine learning (Baro et al., 2015). No matter how good the ML algorithm is, the outcome can be only as good as the veracity of the data feeding the algorithm. Along with variability (see earlier) veracity, i.e., accuracy of data is responsible for the low reproducibility of biomedical, including TBI studies. If clinical decision making in the NICU is automated, based on ML but the data feeding the algorithm is low veracity, the decisions can be life threatening. In experimental TBI, subjectivity, insufficient separation of significant from insignificant and manipulation of primary data can negatively affect the veracity of data.

Validity: Along with veracity, the degree of accuracy and correctness (validity) of data is critical in both experimental and clinical TBI research (Wasser et al., 2015; Belanger et al., 2012; Gardner et al., 2012; Guzel et al., 2010). Primary, raw experimental data are almost always messy and can be inaccurate to a varying degree due to high levels of noise. Establishing the validity of experimental data is critical; primary data need to be cleansed, typically by using rigorous statistical tools. Establishing the validity of data in clinical TBI studies is even more challenging because conditions are less controlled and there can be many confounding factors that may or may not be taken into account.

Vulnerability: Data collected in experimental and clinical TBI studies are vulnerable to manipulation. Experimental data are vulnerable to omissions, willingly or unwillingly removing critical information (Ferguson et al., 2014; Wallis et al., 2013). TBI data can be breached, just like other data, but hacking typically occurs only if there is a commercial or financial interest associated with data.

Volatility: In contrast to, e.g., trying to determine market trends using BD approaches where the key is to use data collected during a predetermined time frame (after that data can be discarded), biomedical, including TBI, research builds on the accumulation of data from which our knowledge derives (Wang and Krishnan, 2014; Sebaa et al., 2018; Luo et al., 2016). Accordingly, TBI research data should be stored indefinitively, so data will be available for future (re) analysis. An important caveat is how useful "old" data are when they are collected by using outdated methods or instruments. Keeping old data can degrade the veracity validity of TBI data and its continuing storage may further complicate data retrieval and (re)analysis.

Visualization: Once BD from experimental and/or clinical TBI research has been processed, it needs to be presented so it is readable and understandable for humans (Dinov, 2016b; Dinov et al., 2016). The classical x and y (maybe also z) variables will not suffice as there are so many variables. Similar to the entire BD field, visualizing and presenting findings from BD analysis of TBI research remain a substantial challenge. Existing tools that include Google Chart, Tableau, Nanocubes, among others and the offers increase weekly.

Value: The potential value of available—legacy—and incoming experimental and clinical TBI data are potentially huge, but it depends on the intended purpose of using BD (Dinov, 2016a; Janke et al., 2016; Luo, 2016). Unless this is clearly defined, the existence of BD alone cannot guarantee that there is a value in the available experimental and clinical TBI data. While there are lots of talk about potential savings in healthcare, including TBI-related costs using BD approaches, these benefits greatly depend on the value of the available data.

5. MACHINE LEARNING

Machine learning (ML), as defined by Arthur Samuel (1959), is a "field of computer science that uses statistical techniques to give computer systems the ability to "learn" with data, without being explicitly programmed" (Mooney and Pejaver, 2017; Helmstaedter, 2015; Liu and Salinas, 2017). ML has evolved from pattern recognition and in contrast to the traditional static program instructions, ML is dynamic, i.e., the algorithm can "learn" from data and can make predictions based on data. ML requires high-speed computing to deal with ever-increasing amounts of data, e.g., to be able to "self-drive" a car (Tesla, Google, Uber), or personalized "recommendations" from Amazon. ML includes various stages or learning tasks, starting with "supervised learning" when a "teacher" presents inputs and expected outputs so the machine can learn the general rule, and ending with unsupervised learning, the machine is left to itself to discover patterns and learn them. The greatest example of machine learning (and artificial intelligence) is AlphaGo, the program developed by Google's DeepMind, which beat the several professional Go players. Go is considered much more difficult for a computer to learn it because, among other issues the much larger branching factor than chess. Branching factor is the number of "choices" at each position the player can take. It is about 35 moves in chess but 250 (average) in Go. The latest version, the AlphaZero algorithm, operates without human data and can learn and master Go in 24 h (!) not only to beat any human in Go but also in chess and in shogi (Japanese chess).

6. ARTIFICIAL INTELLIGENCE

The term artificial intelligence (AI) was introduced by John McCarthy, a cognitive scientist who is called the father of AI. In 1956 Dr. McCarthy coined the term for the first AI conference held at Dartmouth College in 1956. Dr. McCarthy believed that "every aspect of learning or any other feature of intelligence can in principle be so precisely described that a machine can be made to simulate it." But Dr. McCarthy also stated that the breakthroughs needed to make AI a reality can come in "5 or 500 years" (Cascianelli et al., 2016; Chappell and Hawes, 2012; Hassabis et al., 2017).

The term AI has been generally used when a computer "mimics" functions that we humans associate with the human minds, for example learning, problem solving, and decision making.

Today's AI however is rather narrow, only can perform narrow tasks, e.g., speech recognition or facial recognition. The goal is to develop general AI, which can perform

ALL the human tasks and such a general AI will outperform humans in all the tasks. AI uses algorithms, set of unambiguous instructions that a mechanical computer can execute. AI uses highly complex algorithms that built on top of other simpler ones. AI algorithms can learn from data; they can enhance themselves by learning new strategies that have worked well in the past (positive reinforcement). They can also write or modify algorithms.

The capacity of AI greatly depends on machine learning (ML) algorithms, self-learning programs that act like neuronal networks. Like real neuronal networks these are arranged in multiple layers, each sending its output to the next layer and so on. The network of computers with many-many layers can learn highly complex task through so-called deep learning, i.e., each layer "learns" from the other layer and so discover and learn about relationships. Neuronal networks "learn" without specifically being programmed for relationships or rules. The next phase is called "training," machines are fed with curated and labeled data that can be images, texts, etc. The networks "learn" to identify relationships between the labels and, e.g., images. Next step is to provide the machines with "learned," e.g., images mixed with unknown objects or texts, so the network can distinguish between, e.g., the objects or texts, etc. it was trained for and develop relationships or inference (Hassabis et al., 2017; Ghahramani, 2015; Lee et al., 2017).

7. TEXT MINING

Text mining is the technique that extracts high-quality information from texts (Cai et al., 2016; Pasche et al., 2009). The most advanced text mining software use Natural Language Processing (NLP) algorithms. NLP is an AI-based technique that turns unstructured text into a set of features for machine learning to use. The technique using linguistic patterns and terminologies with high precision and ideally requires no manual annotation. NLP recognizes similar concepts, in different contexts and in different grammatical constructions. NLP is using ML techniques and related to AI. NLP can also help ML by enabling to extract a much greater and broader evidence-based and structured data for ML algorithms so they can "learn" from it (Cai et al., 2016; Ruch, 2009).

ML generally requires well-curated input data to generate good results and not unstructured text. NLP can extract clean structured data from legacy publications and reports that are needed to help developing predictive ML NLP algorithms can recognize patterns and trends within a given text (in the given language) and so computers can extract the information required in the format they "understand" and so they can also learn. Importantly, TM extracts precise information; you can search not only for keywords but for phrases, sentences, and relationships.

A special challenge is text mining because texts contain irregularities and ambiguities (Pastur-Romay et al., 2016; Pons et al., 2016; Ruch et al., 2008; Teodoro et al., 2011;

Yadav et al., 2016). A combination of text mining, image, nucleotide, and/or amino acid sequence analysis, and other preprocessing steps are needed to give structure to the raw data and to extract the information or generate quantitative signature vectors enabling data analysis. The most challenging is text preprocessing, requiring statistical parsing, computational linguistics and/or machine learning to generate numerical summaries.

8. EXAMPLES OF USING BD APPROACHES IN TBI RESEARCH

BD is not a panacea, a "solve-it-all" approach. The power of BD is to identify correlations and trends. However, while BD can reveal only correlations but not causations, over time accumulated correlations can allow establishing causative relationships. The successful use in TBI critically depends on three factors: first and foremost, precisely formulated research question(s); availability of data that can be "mined"; availability of algorithm(s) to analyze BD.

Some of the most vexing issues in TBI research can be addressed by finding correlations and establishing trends using BD. Examples include: what are the correlations between injury severity and outcome; between the type of injury and the biological response (secondary injury process); between functional impairments and structural changes; between structural and molecular changes; etc. One of the most important question is, what's the correlation/relationship between experimental TBI data and clinical TBI data. Experimental (preclinical) and clinical TBI studies use different outcome measures, methodologies, different species with different anatomy, physiology, different physical forces resulting in an enormous gap between the two fields. BD approaches have the potential to establish correlations between experimental and clinical TBI data but there are several critical and major hurdles. Using BD in TBI requires first and foremost the availability of Big Data. There are only a few outcome measures, imaging, neurocritical care monitoring, and probably biochemical markers that have been generating and archiving large data sets for potential for DB approaches. In addition, legacy data, peer-reviewed articles, scientific reports have the size for potential use of BD approach for analysis.

9. IMAGING

Neuroimaging, using various modalities, is currently the most promising field to use BD approaches in TBI research, diagnosis, and management (Choudhury et al., 2014; Peng et al., 2016; Kansagra et al., 2016; Webb-Vargas et al., 2017; Toga and Dinov, 2015; Luo et al., 2016; Yadav et al., 2016; Alberich-Bayarri et al., 2017; Kang et al., 2016; Liebeskind et al., 2015; Morris et al., 2018; Smith and Nichols, 2018). Neuroimaging is data rich, data are deposited and archived, and available for analysis using various BD approaches, ML, AI. Advances in pattern recognition that has already resulted in impressive use of ML in medical diagnosis (dermatology) make BD approaches in

TBI imaging closer to reality. BD in neuroimaging however exemplifies also the challenges of using BD in TBI. Neuroimaging in TBI generates (very) large volumes of data; thousands of CT scan and MRI are performed every day following head injuries of different types and intensities. However, unlike other major CNS disorders, Alzheimer's Disease (AD), and stroke, no TBI imaging repository exists. Imaging repositories for AD, Alzheimer's disease Network Initiative ADNI (Hendrix et al., 2015; Toga and Crawford, 2015; Weiner et al., 2015) and for stroke StrokeNet (Liebeskind et al., 2015), are designed with BD approaches in mind.

Analyzing imaging data derived from different institutions is still challenging due to several factors. These include the data format. While DICOM (Digital Imaging and Communications in Medicine) is widely accepted and used native format in which images were acquired, relevant annotations are different and depend on the vendor or the institutions. In the absence of standardized atlas and annotations, the value of the imaging data for BD approaches is rather reduced. Creating a uniform annotation for neuroimaging is critical as it would enable the use of ML to analyze imaging data in the context of other relevant information—available in textual or in structured format—such as medical reports, patient information, laboratory data, etc. Successful creation and use of such annotation has been demonstrated in stroke (Owolabi et al., 2017), but not in TBI. Neuroimaging yields huge amounts of data and the data are stored. CT scans and increasingly MRI are routine after head injury at emergency rooms. However, analyzing imaging data derived from different institutions is challenging due to several factors. These include data format is stored. While there is a widely accepted and used format called DICOM (Digital Imaging and Communications in Medicine), there is also a native format in which images were acquired, relevant annotations are different, and depend on users and vendor. In the absence of standardized annotations, the value of the imaging data is greatly diminished. Creating a uniform annotation for neuroimaging is critical as it would enable the use of ML to analyze imaging data in the context of other relevant information—available in textual or in structured format—such as medical reports, patient information, laboratory data, etc. Successful creation and use of such annotation has been demonstrated in stroke (Owolabi et al., 2017), but not in TBI. Another problem with TBI is that CT and various modalities of MRI (T1- and T2-weighted, fluid attenuated inversion recovery, diffusion- or susceptibility-weighted sequences) can detect acute intracranial sequelae after severe and moderate TBI, but not for the milder form of TBI (Eierud et al., 2014; Liebeskind et al., 2015).

An additional challenge for BD is strangely due to the volume of imaging data. Just the cleaned, annotated published neuroimaging data average 20GB per study. This amount is only the fraction of the total data, publications typically do not include "raw" data. Access to the entire data generated along with annotations would be required for BD approaches. An average fMRI scan can obtain multiple BOLD image volumes of the whole human head per second; advanced DTI imaging is capable of resolving 512 or more fiber directions resulting in increasingly higher spatial resolution and data volume.

These and other technical improvements mean that the amount of data is doubling in roughly every 2 years or faster. Still, dealing with large volume of data is increasingly less of an issue due to increasing processing power than the lack of standardization. Identical CT scanners or MRI machines can generate significantly different quality of primary data and in the absence of standardized annotation, different interpretations. There are different atlases for the normal human brain and different references for disease-related changes. To be able to take full advantage of BD approaches in TBI research will depend on imaging databases that are accessible for investigators and data scientists with full annotations—clinical/medical records, other biomarkers, biochemical, physiological, behavioral, etc. Only if these will be implemented can the TBI field start productively using BD approaches. The amount of data from TBI studies will increase exponentially as more and more institutions are using scanners at increasing frequency and hopefully making the data available. Analyzing imaging data to find correlations between structural and molecular changes (biomarkers) using BD approaches represents an especially serious challenge due to the current issues listed and the varying structures of data. Despite attempts toward structured reporting format in imaging using standardized terminology, the overwhelming majority of data remains unstructured and text heavy. Because of the volume, "manual" curation and interpretation of data is unrealistic, but natural language processing (NLP) offers potential solution. NLP is a computer algorithm that can "translate" human language into a structured data format. Such structured format then can be easily queried and analyzed by existing computer programs.

A successful example of using NLP was reported in pediatric TBI. Yadav et al. (2016) have successfully used free text CT imaging reports for outcome classification. The authors have used an earlier developed combination of NLP and ML for automated classification using the Medical Language Extraction and Encoding (MedLEE; Columbia University, New York, NY; and Health Fidelity, Menlo Park, CA) as NLP platform. The study has used radiological reports as PDF documents. Using optical character recognition, text was extracted from the PDF files and entered into MedLEE. Using half of the reports, MedLEE was trained to create a structured output with standardized medical terms and modifiers. Data were then analyzed using identifiers from the NINDS Common Data Elements (CDE) and analyzed using a decision tree classifier. This approach has resulted in high-fidelity classification, however, with trends toward false positives. Importantly, the current limitations of using this approach seem to be less in the availability of data or NLP and ML technologies, but in the still relatively poor definitions and criteria in the CDE and other databases.

10. BIOCHEMICAL MARKERS

Biochemical markers represent another promising field for using BD approaches. Injury-induced changes in the CSF and serum levels of biochemical markers, especially proteins, have long held the promise to objectively assess injury severity, help determining disease trajectory, predict outcome, and can identify the molecular pathomechanisms (Agoston and Elsayed, 2012; Agoston et al., 2017; Bogoslovsky et al., 2016; Daoud et al., 2014; Kobeissy et al., 2008; Mondello et al., 2017; Strathmann et al., 2014; Wang et al., 2005, 2018; Zetterberg and Blennow, 2016; Zetterberg et al., 2013). Biochemical markers in a clinical setting are measured and recorded in the context of other clinically relevant data—the extent of functional impairments, structural changes obtained by imaging, etc. These data can serve as the critical "training" data for ML/deep learning. Importantly, biomarker data are numeric, typically structured, frequently measured at various postinjury time points, collected in both experimental and in clinical TBI studies, recorded, and stored either in the format of medical reports or publications. Moreover, the biology—cellular origin, cellular functions, etc. of blood and CSF-based biochemical markers especially proteins—has been fairly well established. Also, systems biology enables a much wider utilization of biomarker data as it can reveal correlations between injury-induced changes in biomarker levels and pathological processes.

"Training" deep neuronal networks with data obtained from clinical biomarker monitoring and systems biology would enable to develop inference between, e.g., elevated levels of chemo and cytokines in the CSF and neuroinflammation. Or, "training" the network with biomarker data and protein-protein interaction can "infer" complex molecular pathways that can be used for drug development. In this "training" model, the neuronal network is being fed with condition-specific (e.g., severity level or post-TBI time point), e.g., proteomics data, and the specific functions of the proteins or protein-protein interactions that would allow the network to infer the patterns that is related to, e.g., injury-type specific pathologies. For example, to distinguish between molecular patterns that correlate with the development of cerebral edema, or axonal injury, of inflammation, etc. Another application is to predict long-term outcomes after TBI, such as epilepsy or psychological disorders based on early biomarker data.

There are several challenges using current, existing AI for biomarker-based diagnosis and prognosis in TBI. The main one is that despite their potential and large body of literature reporting biomarker data in TBI, there is no biomarker data repository, so using BD approaches is more challenging than in neuroimaging. At a more technical methodological level, differences in study design and interpretation have made it difficult to validate TBI specific markers. Analytical platforms also vary along with sample preparation, i.e., there are currently no standards or industry-style ISOOs, so biomarker values can (widely) vary from laboratory to laboratory for seemingly identical subjects. Using standardized tests or analytical platforms would be a critical step toward generating reliable and reproducible data that can be used for "training" the machines.

Another issue BD approaches can address is eliminating or reducing investigator bias. As an example, a study analyzing metadata to examine the magnitude of the effect sizes of biomarkers showed that highly cited individual biomarker studies report substantially larger effect estimates for hypothesized associations between injury and biomarker changes than found by the BD approach used to evaluate the very same data (Ioannidis and Panagiotou, 2011).

A critical step to address this issue has been undertaken by the National Institute of Neurological Disorders and Stroke (NINDS) and the Department of Defense (DoD) and created a national resource for archiving and sharing clinical research data on TBI. The Federal Interagency Traumatic Brain Injury Research Informatics System (FITBIR; fitbir.nih.gov) enables data sharing in the field of TBI among individual laboratories. FITBIR currently stores several hundred thousands of data including biomarker data along with demographics, outcome assessments, and imaging. FITBIR provides critical resources and also tools toward utilizing BD approaches in TBI. FITBIR—and CDE—provides detailed procedures for blood and CSF collection in order to promote standardization aimed to improve data quality.

The Common Data Element (CDE) initiative for TBI studies is giant step toward improving data quality By defining disease severity classification, unifying data entry, depositing, and archiving data, CDE can vastly improve data quality and also quantity over time. The success of the initiative is reflected in the increasing entries as well as analyses and studies using FITBIR. Another aspect of efficient use of available biomarker data for ML is to look beyond TBI. TBI shares many neuropathological features, pathomechanism such as inflammation and cellular damage with other CNS diseases and disorders. These databases can be and should be used, provided appropriate criteria, limitations, and restrictions are established from the clinical perspective.

11. LEGACY DATA

Probably the quickest—but not necessarily the easiest—area to demonstrate the enormous potential of BD in TBI is to use existing, legacy data that have accumulated from decades of TBI research, both experimental and clinical (Devine and Zafonte, 2009; Krainin et al., 2011). Analyzing legacy data using BD approaches, AI, ML, can identify critical correlations that humans cannot because of the volume, complexity, and incompleteness of legacy data (Ferguson et al., 2014; Wallis et al., 2013). The main tool needed for successful utilization of BD in analyzing TBI legacy data is text mining (TM). TM is to convert the unstructured information of, e.g., scientific reports into numeric indices that can be stored in relational databases and queried using traditional methods or use the information to "train" neuronal networks in ML. TM can generate data in formats that computer algorithms can "understand," so neuronal networks can be "trained". The key software tool for TM is a computational algorithm, Natural Language Processing (NLP), a form—and the result—of AI. NLP "reads" and interprets the meaning of texts, extracts facts, and importantly establishes relationships. NLP allows to, e.g., recognize similar words and also concepts that may be written differently and—importantly—even if they are used in different contexts. It can also understand analogous expressions as you may have experienced by using, e.g., Google Translate. In addition to power ML, NLP also allows direct mining of existing texts in a systematic way and query textual data for specific information. TM can transform unstructured, textual information into structured, e.g., relational data format that can be analyzed using more traditional approaches or can be integrated with structured data in databases and can be used to "feed or train" ML systems. The main advantage of using TM is that it can perform much more specific searches than current keywords-based search, which retrieves all the documents that contain the keyword(s), e.g., Medline search. The individual still needs to read ALL of the retrieved documents to find the document(s) containing the specific information. TM on the other hand reads the document for you and understands the real meaning using NPL. So, the search using TM will identify not only facts but also relationships between facts, which you use when using the traditional Medline search and can only extract by reading all the retrieved articles and extract data and their relationships. NLP along with ML parts of AI as ML is used to enhance the accuracy of NLP; think about Google translate and compare its accuracy now vs. a few years ago. The (dramatically) increased accuracy is the result of the ongoing ML Google is using to improve the program. On the other side, NPL can also "train" ML by "extracting" a large evidence base of structured data to "feed" the neuronal networks and thus ML can "learn" and becoming more and more precise.

The caveat is that ML requires curated, ideally well curated input data for "training." The quality of input data is the most important determinant of how well ML will perform its intended tasks. Translated this into "mining" legacy TBI data, it means that published articles should be "readable" by ML through NLP, so relationships and correlations can be accurately established.

A major hindrance using BD approaches to "mine" legacy TBI data is that the published data represent a fraction of the data collected (Ferguson et al., 2014; Wallis et al., 2013). If you plot all data collected during a study—clinical or experimental—so called "long tail" of data, the published data represent—at most— ~50% of ALL data collected. In addition, data submitted for publication are "curated," may be tailored to support the hypothesis. Massive amount of critical data is never "digitized," the "dark data" represent handwritten laboratory and clinical notes, animal care records, etc. If it is in digital format, significant proportion of data is stored on personal hard drives ("file-drawer phenomenon"). The current total TBI literature published in peer-reviewed journals is ~40,000 papers. Access (digital) to "dark data" would at least double the number. This number sounds big but may not be big enough for BD as the legacy data are fragmented into subtopics and within a subtopic the data volume is not sufficient. This problem has been illustrated in a study by Bennett et al. (Bennett et al., 2017) and reviewed by Horvat and Kochanek (2017). In the original study (Bennett et al., 2017), the authors used ML to analyze data in two linked national databases with data from 30 US children's hospitals on 3084 children with severe TBI to determine the correlation between intracranial pressure monitoring and outcome. While the subject number—more than 3000—looks a lot, data are still not BIG enough to generate definitive correlation between ICP monitoring and outcome. Patient heterogeneity in TBI is known to be extremely high, resulting in large numbers of covariants and accordingly in a huge complexity of symptoms; thus, as the authors concluded if a much, much larger data set is needed, they recommend a large prospective cohort study or randomized trial to successfully use BD approaches to address fundamental relationships (Bennett et al., 2017). As summarized in the accompanying Editorial, "Big Data is not yet big enough…" (Horvat and Kochanek, 2017).

12. FUTURE OF BIG DATA IN TBI

The potential value of using BD approaches in healthcare—including TBI—is huge. Conservative estimates put the saving by BD with currently (!) available technologies as ~15% of healthcare costs. Given that the DIRECT medical, economic, and social expenses related to TBI are ~ 96 billion dollars EVERY year in the US, the saving would be ~15 billion dollars per year. TBI increases the risk of developing chronic neurodegenerative conditions, Alzheimer's disease, and/or chronic traumatic encephalopathy (CTE); thus, the savings can be substantially higher. The reason why BD approaches have not been successfully used in TBI is simple; there are no sufficiently BIG data available to utilize ML, AI, and other BD tools. The tools are there and getting better and better by the day. As the aforementioned study (Bennett et al., 2017) and Editorial (Horvat and Kochanek, 2017) illustrate, data volume in TBI is still not BIG enough to use the power of BD approaches to find relationship between even relatively straightforward and relatively well-documented variants, ICP monitoring and outcomes. Given the heterogeneity of TBI patients and the complexity of pathological changes, using BD approaches in TBI must have BIG, very, very BIG data. The sheer number of potential covariants in clinical or even experimental TBI is enormous, and currently large portion of the data generated is not recorded or not available digitally.

Recommendations: Data, data, data…. Large volumes of incomplete messy data are more valuable than clean, curated but small data sets. Smaller sets may or may not be representative and almost certainly biased due to "curation." Call the messy approach Google approach. Google Translate—you are probably using it—is based on massive amounts of—messy—data. Google collected—vacuumed up—all available data from the Internet, as messy, incomplete, and inaccurate source of all the languages—currently 100 plus. Using Phrase-Based Machine Translation (PBMT) and Neural Machine Translation (NMT), one can translate between these languages. Moreover, it can translate 32 languages via voice in "conversation mode," and 27 languages via real-time video in "augmented reality mode." What made these amazing "tools" possible is the volume of available data so ML and AI could have been developed. The 100% failure rate of clinical trials indicates that more of the same will not work or in other words, TBI research has reached its "strategic inflection point." The field urgently needs change, which mostly entails to changing habits; (1) record and store EVERYTHING. Databases like FITBIR using CDE is a great start. Encouragement or mandating researchers and clinicians receiving public money for their TBI research to record and enter all data into a public database; this will lead to reduction and/or elimination of "dark data"; (2) make ALL data available, like FITBIR is a good start and great example; (3) start using existing ML, AI NLP, etc. tools to analyze data, so incomplete, unformatted, fragmented, unstructured data—including legacy data. By mining existing data for new information, relationships and correlations will teach us how current tools need to be modified or tailored for TBI and how data need to be entered and what are the minimum requirements for data quality.

REFERENCES

Addona, T.A., Abbatiello, S.E., Schilling, B., Skates, S.J., Mani, D.R., Bunk, D.M., Spiegelman, C.H., Zimmerman, L.J., Ham, A.J., Keshishian, H., Hall, S.C., Allen, S., Blackman, R.K., Borchers, C.H., Buck, C., Cardasis, H.L., Cusack, M.P., Dodder, N.G., Gibson, B.W., Held, J.M., Hiltke, T., Jackson, A., Johansen, E.B., Kinsinger, C.R., Li, J., Mesri, M., Neubert, T.A., Niles, R.K., Pulsipher, T.C., Ransohoff, D., Rodriguez, H., Rudnick, P.A., Smith, D., Tabb, D.L., Tegeler, T.J., Variyath, A.M., Vega-Montoto, L.J., Wahlander, A., Waldemarson, S., Wang, M., Whiteaker, J.R., Zhao, L., Anderson, N.L., Fisher, S.J., Liebler, D.C., Paulovich, A.G., Regnier, F.E., Tempst, P., Carr, S.A., 2009. Multi-site assessment of the precision and reproducibility of multiple reaction monitoring-based measurements of proteins in plasma. Nat. Biotechnol. 27, 633–641.
Agoston, D., 2015. Bench-to-bedside and bedside back to the bench: seeking a better understanding of the acute pathophysiological process in severe traumatic brain injury. Front. Neurol 6, 1–6.
Agoston, D.V., 2017. Understanding the complexities of traumatic brain injury: a big data approach to a big disease. Behav. Brain Res 340, 172–173.
Agoston, D.V., Elsayed, M., 2012. Serum-based protein biomarkers in blast-induced traumatic brain injury spectrum disorder. Front. Neurol. 3, 107.
Agoston, D.V., Kamnaksh, A., 2015. Modeling the neurobehavioral consequences of blast-induced traumatic brain injury spectrum disorder and identifying related biomarkers. In: Kobeissy, F.H. (Ed.), Brain Neurotrauma: Molecular, Neuropsychological, and Rehabilitation Aspects. CRC Press/Taylor & Francis, Boca Raton, FL.
Agoston, D.V., Langford, T.D., 2017. Big Data in Traumatic Brain Injury. Concussion 3, 45–59.
Agoston, D.V., Shutes-David, A., Peskind, E.R., 2017. Biofluid biomarkers of traumatic brain injury. Brain Inj. 31, 1195–1203.
Alberich-Bayarri, A., Hernandez-Navarro, R., Ruiz-Martinez, E., Garcia-Castro, F., Garcia-Juan, D., Marti-Bonmati, L., 2017. Development of imaging biomarkers and generation of big data. Radiol. Med. 122, 444–448.
Andreu-Perez, J., Poon, C.C., Merrifield, R.D., Wong, S.T., Yang, G.Z., 2015. Big data for health. IEEE J. Biomed. Health Inform. 19, 1193–1208.
Andriessen, T.M., Horn, J., Franschman, G., van der Naalt, J., Haitsma, I., Jacobs, B., Steyerberg, E.W., Vos, P.E., 2011. Epidemiology, severity classification, and outcome of moderate and severe traumatic brain injury: a prospective multicenter study. J. Neurotrauma 28, 2019–2031.

Baro, E., Degoul, S., Beuscart, R., Chazard, E., 2015. Toward a literature-driven definition of big data in healthcare. Biomed. Res. Int. 2015, 639021.

Belanger, H.G., Vanderploeg, R.D., Soble, J.R., Richardson, M., Groer, S., 2012. Validity of the Veterans Health Administration's traumatic brain injury screen. Arch. Phys. Med. Rehabil 28, 5–12.

Bell, R.S., Ecker, R.D., Severson 3rd, M.A., Wanebo, J.E., Crandall, B., Armonda, R.A., 2010. The evolution of the treatment of traumatic cerebrovascular injury during wartime. Neurosurg. Focus 28, E5.

Bennett, T.D., DeWitt, P.E., Greene, T.H., Srivastava, R., Riva-Cambrin, J., Nance, M.L., Bratton, S.L., Runyan, D.K., Dean, J.M., Keenan, H.T., 2017. Functional outcome after intracranial pressure monitoring for children with severe traumatic brain injury. JAMA Pediatr. 171, 965–971.

Bergsneider, M., Hovda, D.A., Lee, S.M., Kelly, D.F., McArthur, D.L., Vespa, P.M., Lee, J.H., Huang, S.C., Martin, N.A., Phelps, M.E., Becker, D.P., 2000. Dissociation of cerebral glucose metabolism and level of consciousness during the period of metabolic depression following human traumatic brain injury. J. Neurotrauma 17, 389–401.

Bergsneider, M., Hovda, D.A., McArthur, D.L., Etchepare, M., Huang, S.C., Sehati, N., Satz, P., Phelps, M.E., Becker, D.P., 2001. Metabolic recovery following human traumatic brain injury based on FDG-PET: time course and relationship to neurological disability. J. Head Trauma Rehabil. 16, 135–148.

Bigler, E.D., 2001. Quantitative magnetic resonance imaging in traumatic brain injury. J. Head Trauma Rehabil. 16, 117–134.

Bogoslovsky, T., Gill, J., Jeromin, A., Davis, C., Diaz-Arrastia, R., 2016. Fluid biomarkers of traumatic brain injury and intended context of use. Diagnostics, 6, 4–11.

Bolouri, H., Zhao, L.P., Holland, E.C., 2016. Big data visualization identifies the multidimensional molecular landscape of human gliomas. Proc. Natl. Acad. Sci. USA 113, 5394–5399.

Bouzat, P., Oddo, M., Payen, J.F., 2014. Transcranial Doppler after traumatic brain injury: is there a role? Curr. Opin. Crit. Care 20, 153–160.

Browne, K.D., Chen, X.H., Meaney, D.F., Smith, D.H., 2011. Mild traumatic brain injury and diffuse axonal injury in swine. J. Neurotrauma 28, 1747–1755.

Buki, A., Povlishock, J.T., 2006. All roads lead to disconnection?—Traumatic axonal injury revisited. Acta Neurochir. (Wien) 148, 181–193 (discussion 193–184).

Bzdok, D., Yeo, B.T.T., 2017. Inference in the age of big data: future perspectives on neuroscience. Neuroimage 155, 549–564.

Cahsai, A., Anagnostopoulos, C., Triantafillou, P., 2015. Scalable data quality for big data: the pythia framework for handling missing values. Big Data 3, 159–172.

Cai, T., Giannopoulos, A.A., Yu, S., Kelil, T., Ripley, B., Kumamaru, K.K., Rybicki, F.J., Mitsouras, D., 2016. Natural language processing technologies in radiology research and clinical applications. Radiographics 36, 176–191.

Cascianelli, S., Scialpi, M., Amici, S., Forini, N., Minestrini, M., Fravolini, M.L., Sinzinger, H., Schillaci, O., Palumbo, B., 2016. Role of artificial intelligence techniques (automatic classifiers) in molecular imaging modalities in neurodegenerative diseases. Curr. Alzheimer Res 14, 198–207.

Chappell, J., Hawes, N., 2012. Biological and artificial cognition: what can we learn about mechanisms by modelling physical cognition problems using artificial intelligence planning techniques? Philos. Trans. R. Soc. Lond. B Biol. Sci. 367, 2723–2732.

Chen, Y., Elenee Argentinis, J.D., Weber, G., 2016. IBM watson: how cognitive computing can be applied to big data challenges in life sciences research. Clin. Ther. 38, 688–701.

Choudhury, S., Fishman, J.R., McGowan, M.L., Juengst, E.T., 2014. Big data, open science and the brain: lessons learned from genomics. Front. Hum. Neurosci. 8, 239.

Daoud, H., Alharfi, I., Alhelali, I., Charyk Stewart, T., Qasem, H., Fraser, D.D., 2014. Brain injury biomarkers as outcome predictors in pediatric severe traumatic brain injury. Neurocrit. Care 20, 427–435.

De Guio, F., Jouvent, E., Biessels, G.J., Black, S.E., Brayne, C., Chen, C., Cordonnier, C., De Leeuw, F.E., Dichgans, M., Doubal, F., Duering, M., Dufouil, C., Duzel, E., Fazekas, F., Hachinski, V., Ikram, M.A., Linn, J., Matthews, P.M., Mazoyer, B., Mok, V., Norrving, B., O'Brien, J.T., Pantoni, L., Ropele, S., Sachdev, P., Schmidt, R., Seshadri, S., Smith, E.E., Sposato, L.A., Stephan, B., Swartz, R.H., Tzourio, C., van Buchem, M., van der Lugt, A., van Oostenbrugge, R., Vernooij, M.W.,

Viswanathan, A., Werring, D., Wollenweber, F., Wardlaw, J.M., Chabriat, H., 2016. Reproducibility and variability of quantitative magnetic resonance imaging markers in cerebral small vessel disease. J. Cereb. Blood Flow Metab. 36, 1319–1337.

Devine, J.M., Zafonte, R.D., 2009. Physical exercise and cognitive recovery in acquired brain injury: a review of the literature. PM R. 1, 560–575.

Devinsky, O., Dilley, C., Ozery-Flato, M., Aharonov, R., Goldschmidt, Y., Rosen-Zvi, M., Clark, C., Fritz, P., 2016. Changing the approach to treatment choice in epilepsy using big data. Epilepsy Behav. 56, 32–37.

Dilsizian, S.E., Siegel, E.L., 2014. Artificial intelligence in medicine and cardiac imaging: harnessing big data and advanced computing to provide personalized medical diagnosis and treatment. Curr. Cardiol. Rep. 16, 441.

Dinov, I.D., 2016a. Volume and value of big healthcare data. J. Med. Stat. Inform. 4.

Dinov, I.D., 2016b. Methodological challenges and analytic opportunities for modeling and interpreting Big Healthcare Data. GigaScience 5, 12.

Dinov, I.D., Heavner, B., Tang, M., Glusman, G., Chard, K., Darcy, M., Madduri, R., Pa, J., Spino, C., Kesselman, C., Foster, I., Deutsch, E.W., Price, N.D., Van Horn, J.D., Ames, J., Clark, K., Hood, L., Hampstead, B.M., Dauer, W., Toga, A.W., 2016. Predictive big data analytics: a study of Parkinson's disease using large, complex, heterogeneous, incongruent, multi-source and incomplete observations. PLoS One 11, e0157077.

Duhaime, A.C., Beckwith, J.G., Maerlender, A.C., McAllister, T.W., Crisco, J.J., Duma, S.M., Brolinson, P.G., Rowson, S., Flashman, L.A., Chu, J.J., Greenwald, R.M., 2012. Spectrum of acute clinical characteristics of diagnosed concussions in college athletes wearing instrumented helmets: clinical article. J. Neurosurg. 117, 1092–1099.

Eierud, C., Craddock, R.C., Fletcher, S., Aulakh, M., King-Casas, B., Kuehl, D., LaConte, S.M., 2014. Neuroimaging after mild traumatic brain injury: review and meta-analysis. NeuroImage. Clinical 4, 283–294.

Esselman, P.C., Uomoto, J.M., 1995. Classification of the spectrum of mild traumatic brain injury. Brain Inj. 9, 417–424.

Ferguson, A.R., Nielson, J.L., Cragin, M.H., Bandrowski, A.E., Martone, M.E., 2014. Big data from small data: data-sharing in the 'long tail' of neuroscience. Nat. Neurosci. 17, 1442–1447.

Finnie, J.W., 2013. Neuroinflammation: beneficial and detrimental effects after traumatic brain injury. Inflammopharmacology 21, 309–320.

Flechet, M., Grandas, F.G., Meyfroidt, G., 2016. Informatics in neurocritical care: new ideas for Big Data. Curr. Opin. Crit. Care 22, 87–93.

Fundamental Neuroscience, fourth ed. Elsevier, Schneider, New York.

Gardner, A., Kay-Lambkin, F., Stanwell, P., Donnelly, J., Williams, W.H., Hiles, A., Schofield, P., Levi, C., Jones, D.K., 2012. A systematic review of diffusion tensor imaging findings in sports-related concussion. J. Neurotrauma 29, 2521–2538.

Ghahramani, Z., 2015. Probabilistic machine learning and artificial intelligence. Nature 521, 452–459.

Golding, E.M., 2002. Sequelae following traumatic brain injury. The cerebrovascular perspective. Brain Res. Brain Res. Rev. 38, 377–388.

Guzel, A., Karasalihoglu, S., Aylanc, H., Temizoz, O., Hicdonmez, T., 2010. Validity of serum tau protein levels in pediatric patients with minor head trauma. Am. J. Emerg. Med. 28, 399–403.

Hasoon, J., 2017. Blast-associated traumatic brain injury in the military as a potential trigger for dementia and chronic traumatic encephalopathy. U.S. Army Med. Dep. J, 102–105.

Hassabis, D., Kumaran, D., Summerfield, C., Botvinick, M., 2017. Neuroscience-inspired artificial intelligence. Neuron 95, 245–258.

Hay, J., Johnson, V.E., Smith, D.H., Stewart, W., 2016. Chronic traumatic encephalopathy: the neuropathological legacy of traumatic brain injury. Annu. Rev. Pathol 11, 21–45.

Helmstaedter, M., 2015. The mutual inspirations of machine learning and neuroscience. Neuron 86, 25–28.

Hendrix, J.A., Finger, B., Weiner, M.W., Frisoni, G.B., Iwatsubo, T., Rowe, C.C., Kim, S.Y., Guinjoan, S.M., Sevlever, G., Carrillo, M.C., 2015. The worldwide Alzheimer's disease neuroimaging initiative: an update. Alzheimers Dement. 11, 850–859.

Hill, C.S., Coleman, M.P., Menon, D.K., 2016. Traumatic axonal injury: mechanisms and translational opportunities. Trends Neurosci. 39, 311–324.

Horvat, C.M., Kochanek, P.M., 2017. Big Data not yet big enough to determine the influence of intracranial pressure monitoring on outcome in children with severe traumatic brain injury. JAMA Pediatr. 171, 942–943.

Ioannidis, J.P., Panagiotou, O.A., 2011. Comparison of effect sizes associated with biomarkers reported in highly cited individual articles and in subsequent meta-analyses. JAMA 305, 2200–2210.

Irimia, A., Wang, B., Aylward, S.R., Prastawa, M.W., Pace, D.F., Gerig, G., Hovda, D.A., Kikinis, R., Vespa, P.M., Van Horn, J.D., 2012. Neuroimaging of structural pathology and connectomics in traumatic brain injury: toward personalized outcome prediction. NeuroImage. Clinical 1, 1–17.

Janke, A.T., Overbeek, D.L., Kocher, K.E., Levy, P.D., 2016. Exploring the potential of predictive analytics and big data in emergency care. Ann. Emerg. Med. 67, 227–236.

Jellinger, K.A., 2004. Traumatic brain injury as a risk factor for Alzheimer's disease. J. Neurol. Neurosurg. Psychiatry 75, 511–512.

Johnson, V.E., Stewart, W., Smith, D.H., 2013. Axonal pathology in traumatic brain injury. Exp. Neurol. 246, 35–43.

Kang, J., Caffo, B., Liu, H., 2016. Editorial: recent advances and challenges on big data analysis in neuroimaging. Front. Neurosci. 10, 505.

Kansagra, A.P., Yu, J.P., Chatterjee, A.R., Lenchik, L., Chow, D.S., Prater, A.B., Yeh, J., Doshi, A.M., Hawkins, C.M., Heilbrun, M.E., Smith, S.E., Oselkin, M., Gupta, P., Ali, S., 2016. Big data and the future of radiology informatics. Acad. Radiol. 23, 30–42.

Kiernan, P.T., Montenigro, P.H., Solomon, T.M., McKee, A.C., 2015. Chronic traumatic encephalopathy: a neurodegenerative consequence of repetitive traumatic brain injury. Semin. Neurol. 35, 20–28.

Kissin, I., 2018. What can big data on academic interest reveal about a drug? Reflections in three major US databases. Trends Pharmacol. Sci. 39, 248–257.

Kobeissy, F.H., Sadasivan, S., Oli, M.W., Robinson, G., Larner, S.F., Zhang, Z., Hayes, R.L., Wang, K.K., 2008. Neuroproteomics and systems biology-based discovery of protein biomarkers for traumatic brain injury and clinical validation. Proteomics Clin. Appl. 2, 1467–1483.

Krainin, B.M., Forsten, R.D., Kotwal, R.S., Lutz, R.H., Guskiewicz, K.M., 2011. Mild traumatic brain injury literature review and proposed changes to classification. J. Spec. Oper. Med. 11, 38–47.

Kumar, A., Loane, D.J., 2012. Neuroinflammation after traumatic brain injury: opportunities for therapeutic intervention. Brain Behav Immun 26, 1191–1201.

Lebo, M.S., Sutti, S., Green, R.C., 2016. Big Data. Gets Personal. Sci. Transl. Med. 8, 322fs323–323fs323.

Lee, E.J., Kim, Y.H., Kim, N., Kang, D.W., 2017. Deep into the brain: artificial intelligence in stroke imaging. J. Stroke 19, 277–285.

Liebeskind, D.S., Albers, G.W., Crawford, K., Derdeyn, C.P., George, M.S., Palesch, Y.Y., Toga, A.W., Warach, S., Zhao, W., Brott, T.G., Sacco, R.L., Khatri, P., Saver, J.L., Cramer, S.C., Wolf, S.L., Broderick, J.P., Wintermark, M., 2015. Imaging in StrokeNet: realizing the potential of big data. Stroke 46, 2000–2006.

Liu, N.T., Salinas, J., 2017. Machine learning for predicting outcomes in trauma. Shock 48, 504–510.

Logsdon, A.F., Lucke-Wold, B.P., Turner, R.C., Huber, J.D., Rosen, C.L., Simpkins, J.W., 2015. Role of microvascular disruption in brain damage from traumatic brain injury. Compr. Physiol. 5, 1147–1160.

Lucke-Wold, B.P., Turner, R.C., Logsdon, A.F., Bailes, J.E., Huber, J.D., Rosen, C.L., 2014. Linking traumatic brain injury to chronic traumatic encephalopathy: identification of potential mechanisms leading to neurofibrillary tangle development. J. Neurotrauma. 31, 1129–1138.

Luo, G., 2016. PredicT-ML: a tool for automating machine learning model building with big clinical data. Health Inf. Sci. Syst. 4, 5.

Luo, J., Wu, M., Gopukumar, D., Zhao, Y., 2016. Big Data application in biomedical research and health care: a literature review. Biomed. Inform. Insights 8, 1–10.

Manley, G.T., Diaz-Arrastia, R., Brophy, M., Engel, D., Goodman, C., Gwinn, K., Veenstra, T.D., Ling, G., Ottens, A.K., Tortella, F., Hayes, R.L., 2010. Common data elements for traumatic brain injury: recommendations from the biospecimens and biomarkers working group. Arch. Phys. Med. Rehabil. 91, 1667–1672.

Manor, T., Barbiro-Michaely, E., Rogatsky, G., Mayevsky, A., 2008. Real-time multi-site multi-parametric monitoring of rat brain subjected to traumatic brain injury. Neurol. Res. 30, 1075–1083.

Matz, P.G., Pitts, L., 1997. Monitoring in traumatic brain injury. Clin. Neurosurg. 44, 267–294.

McIntyre, R.S., Cha, D.S., Jerrell, J.M., Swardfager, W., Kim, R.D., Costa, L.G., Baskaran, A., Soczynska, J.K., Woldeyohannes, H.O., Mansur, R.B., Brietzke, E., Powell, A.M., Gallaugher, A., Kudlow, P., Kaidanovich-Beilin, O., Alsuwaidan, M., 2014. Advancing biomarker research: utilizing 'Big Data' approaches for the characterization and prevention of bipolar disorder. Bipolar Disord. 16, 531–547.

McKee, A.C., Stein, T.D., Nowinski, C.J., Stern, R.A., Daneshvar, D.H., Alvarez, V.E., Lee, H.S., Hall, G., Wojtowicz, S.M., Baugh, C.M., Riley, D.O., Kubilus, C.A., Cormier, K.A., Jacobs, M.A., Martin, B.R., Abraham, C.R., Ikezu, T., Reichard, R.R., Wolozin, B.L., Budson, A.E., Goldstein, L.E., Kowall, N.W., Cantu, R.C., 2013. The spectrum of disease in chronic traumatic encephalopathy. Brain 136, 43–64.

Mondello, S., Sorinola, A., Czeiter, E., Vamos, Z., Amrein, K., Synnot, A., Donoghue, E.L., Sandor, J., Wang, K.K.W., Diaz-Arrastia, R., Steyerberg, E.W., Menon, D., Maas, A., Buki, A., 2017. Blood-based protein biomarkers for the management of traumatic brain injuries in adults presenting with mild head injury to emergency departments: a living systematic review and meta-analysis. J. Neurotrauma 34, 1–21.

Mooney, S.J., Pejaver, V., 2017. Big data in public health: terminology, machine learning, and privacy. Annu. Rev. Public Health 39, 95–112.

Moore, A.H., Osteen, C.L., Chatziioannou, A.F., Hovda, D.A., Cherry, S.R., 2000. Quantitative assessment of longitudinal metabolic changes in vivo after traumatic brain injury in the adult rat using FDG-microPET. J. Cereb. Blood Flow Metab. 20, 1492–1501.

Morris, M.A., Saboury, B., Burkett, B., Gao, J., Siegel, E.L., 2018. Reinventing radiology: big data and the future of medical imaging. J. Thorac. Imaging 33, 4–16.

Mu, W., Catenaccio, E., Lipton, M.L., 2016. Neuroimaging in blast-related mild traumatic brain injury. J. Head Trauma Rehabil 32, 55–69.

Owolabi, M., Ogbole, G., Akinyemi, R., Salaam, K., Akpa, O., Mongkolwat, P., Omisore, A., Agunloye, A., Efidi, R., Odo, J., Makanjuola, A., Akpalu, A., Sarfo, F., Owolabi, L., Obiako, R., Wahab, K., Sanya, E., Adebayo, P., Komolafe, M., Adeoye, A.M., Fawale, M.B., Akinyemi, J., Osaigbovo, G., Sunmonu, T., Olowoyo, P., Chukwuonye, I., Obiabo, Y., Ibinaiye, P., Dambatta, A., Mensah, Y., Abdul, S., Olabinri, E., Ikubor, J., Oyinloye, O., Odunlami, F., Melikam, E., Saulson, R., Kolo, P., Ogunniyi, A., Ovbiagele, B., 2017. Development and reliability of a user-friendly multicenter phenotyping application for hemorrhagic and ischemic stroke. J. Stroke Cerebrovasc. Dis. 26, 2662–2670.

Pasche, E., Teodoro, D., Gobeill, J., Ruch, P., Lovis, C., 2009. Automatic medical knowledge acquisition using question-answering. Stud. Health Technol. Inform. 150, 569–573.

Pastur-Romay, L.A., Cedron, F., Pazos, A., Porto-Pazos, A.B., 2016. Deep artificial neural networks and neuromorphic chips for big data analysis: pharmaceutical and bioinformatics applications. Int. J. Mol. Sci 17, 1313–1320.

Peng, H., Zhou, J., Zhou, Z., Bria, A., Li, Y., Kleissas, D.M., Drenkow, N.G., Long, B., Liu, X., Chen, H., 2016. Bioimage informatics for big data. Adv. Anat. Embryol. Cell. Biol. 219, 263–272.

Pons, E., Braun, L.M., Hunink, M.G., Kors, J.A., 2016. Natural language processing in radiology: a systematic review. Radiology 279, 329–343.

Portbury, S.D., Adlard, P.A., 2015. Traumatic brain injury, chronic traumatic encephalopathy, and Alzheimer's disease: common pathologies potentiated by altered zinc homeostasis. J. Alzheimers Dis. 46, 297–311.

Povlishock, J.T., 1992. Traumatically induced axonal injury: pathogenesis and pathobiological implications. Brain Pathol. 2, 1–12.

Povlishock, J.T., 1993. Pathobiology of traumatically induced axonal injury in animals and man. Ann. Emerg. Med. 22, 980–986.

Povlishock, J.T., Christman, C.W., 1995. The pathobiology of traumatically induced axonal injury in animals and humans: a review of current thoughts. J. Neurotrauma 12, 555–564.

Rodriguez, A., Smielewski, P., Rosenthal, E., Moberg, D., 2018. Medical device connectivity challenges outline the technical requirements and standards for promoting big data research and personalized medicine in neurocritical care. Mil. Med. 183, 99–104.

Ruch, P., 2009. A medical informatics perspective on decision support: toward a unified research paradigm combining biological vs. clinical, empirical vs. legacy, and structured vs. unstructured data. Yearb. Med. Inform, 96–98.

Ruch, P., Gobeilla, J., Tbahritia, I., Geissbuhlera, A., 2008. From episodes of care to diagnosis codes: automatic text categorization for medico-economic encoding. AMIA Annu. Symp. Proc, 636–640.

Sebaa, A., Chikh, F., Nouicer, A., Tari, A., 2018. Medical big data warehouse: architecture and system design, a case study: improving healthcare resources distribution. J. Med. Syst. 42, 59.

Shen, Y., Kou, Z., Kreipke, C.W., Petrov, T., Hu, J., Haacke, E.M., 2007. In vivo measurement of tissue damage, oxygen saturation changes and blood flow changes after experimental traumatic brain injury in rats using susceptibility weighted imaging. Magn. Reson. Imaging 25, 219–227.

Shulman, A., Strashun, A.M., 2009. Fluid dynamics vascular theory of brain and inner-ear function in traumatic brain injury: a translational hypothesis for diagnosis and treatment. Int. Tinnitus J. 15, 119–129.

Siddiqui, J., Grover, P.J., Makalanda, H.L., Campion, T., Bull, J., Adams, A., 2017. The spectrum of traumatic injuries at the craniocervical junction: a review of imaging findings and management. Emerg. Radiol. 24, 377–385.

Simon, D.W., McGeachy, M.J., Bayir, H., Clark, R.S., Loane, D.J., Kochanek, P.M., 2017. The far-reaching scope of neuroinflammation after traumatic brain injury. Nat Rev. Neurol. 13, 171–191.

Sivanandam, T.M., Thakur, M.K., 2012. Traumatic brain injury: a risk factor for Alzheimer's disease. Neurosci. Biobehav. Rev. 36, 1376–1381.

Smith, M., 2008. Monitoring intracranial pressure in traumatic brain injury. Anesth. Analg. 106, 240–248.

Smith, S.M., Nichols, T.E., 2018. Statistical challenges in "big data" human neuroimaging. Neuron 97, 263–268.

Strathmann, F.G., Schulte, S., Goerl, K., Petron, D.J., 2014. Blood-based biomarkers for traumatic brain injury: evaluation of research approaches, available methods and potential utility from the clinician and clinical laboratory perspectives. Clin. Biochem. 47, 876–888.

Talboom, J.S., Huentelman, M.J., 2018. Big data collision: the internet of things, wearable devices, and genomics in the study of neurological traits and disease. Hum. Mol. Genet 27, 35–39.

Teodoro, D., Choquet, R., Schober, D., Mels, G., Pasche, E., Ruch, P., Lovis, C., 2011. Interoperability driven integration of biomedical data sources. Stud. Health Technol. Inform. 169, 185–189.

Terry, D.E., Desiderio, D.M., 2003. Between-gel reproducibility of the human cerebrospinal fluid proteome. Proteomics 3, 1962–1979.

Toga, A.W., Crawford, K.L., 2015. The Alzheimer's disease neuroimaging initiative informatics core: a decade in review. Alzheimers Dement. 11, 832–839.

Toga, A.W., Dinov, I.D., 2015. Sharing big biomedical data. J. Big Data 2, 1–7.

Vallmuur, K., Marucci-Wellman, H.R., Taylor, J.A., Lehto, M., Corns, H.L., Smith, G.S., 2016. Harnessing information from injury narratives in the 'big data' era: understanding and applying machine learning for injury surveillance. Inj. Prev. 22 (Suppl. 1), i34–i42.

Van Horn, J.D., Toga, A.W., 2014. Human neuroimaging as a "Big Data" science. Brain Imaging Behav 8, 323–331.

Vespa, P., Bergsneider, M., Hattori, N., Wu, H.M., Huang, S.C., Martin, N.A., Glenn, T.C., McArthur, D.L., Hovda, D.A., 2005. Metabolic crisis without brain ischemia is common after traumatic brain injury: a combined microdialysis and positron emission tomography study. J. Cereb. Blood Flow Metab. 25, 763–774.

Wallis, J.C., Rolando, E., Borgman, C.L., 2013. If we share data, will anyone use them? Data sharing and reuse in the long tail of science and technology. PLoS One 8, e67332.

Wang, W., Krishnan, E., 2014. Big data and clinicians: a review on the state of the science. JMIR Med. Inform. 2, e1.

Wang, K.K., Ottens, A.K., Liu, M.C., Lewis, S.B., Meegan, C., Oli, M.W., Tortella, F.C., Hayes, R.L., 2005. Proteomic identification of biomarkers of traumatic brain injury. Expert Rev. Proteomics 2, 603–614.

Wang, K.K., Yang, Z., Zhu, T., Shi, Y., Rubenstein, R., Tyndall, J.A., Manley, G.T., 2018. An update on diagnostic and prognostic biomarkers for traumatic brain injury. Expert. Rev. Mol. Diagn. 18, 165–180.

Wasser, T., Haynes, K., Barron, J., Cziraky, M., 2015. Using 'big data' to validate claims made in the pharmaceutical approval process. J. Med. Econ. 18, 1013–1019.

Webb-Vargas, Y., Chen, S., Fisher, A., Mejia, A., Xu, Y., Crainiceanu, C., Caffo, B., Lindquist, M.A., 2017. Big data and neuroimaging. Stat. Biosci. 9, 543–558.

Weiner, M.W., Veitch, D.P., Aisen, P.S., Beckett, L.A., Cairns, N.J., Cedarbaum, J., Donohue, M.C., Green, R.C., Harvey, D., Jack Jr., C.R., Jagust, W., Morris, J.C., Petersen, R.C., Saykin, A.J., Shaw, L., Thompson, P.M., Toga, A.W., Trojanowski, J.Q., 2015. Impact of the Alzheimer's disease neuroimaging initiative, 2004 to 2014. Alzheimers Dement. 11, 865–884.

Wheble, J.L., Menon, D.K., 2016. TBI-the most complex disease in the most complex organ: the CENTER-TBI trial-a commentary. J. R. Army Med. Corps 162, 87–89.

White, T.E., Ford, B.D., 2015. Frontiers in neuroengineering gene interaction hierarchy analysis can be an effective tool for managing big data related to unilateral traumatic brain injury. In: Kobeissy, F.H. (Ed.), Brain Neurotrauma: Molecular, Neuropsychological, and Rehabilitation Aspects. CRC Press/Taylor & Francis, Boca Raton, FL.

Yadav, K., Sarioglu, E., Choi, H.A., Cartwright, W.B., Hinds, P.S., Chamberlain, J.M., 2016. Automated outcome classification of computed tomography imaging reports for pediatric traumatic brain injury. Acad. Emerg. Med. 23, 171–178.

Zetterberg, H., Blennow, K., 2016. Fluid biomarkers for mild traumatic brain injury and related conditions. Nat. Rev. Neurol. 12, 563–574.

Zetterberg, H., Smith, D.H., Blennow, K., 2013. Biomarkers of mild traumatic brain injury in cerebrospinal fluid and blood. Nat. Rev. Neurol 9, 201–210.

CHAPTER 5

Artificial Intelligence Integration for Neurodegenerative Disorders

Rajat Vashistha*,ᵃ, Dinesh Yadav*,ᵃ, Deepak Chhabra*, Pratyoosh Shukla†,**
*Optimization and Mechatronics Laboratory, Department of Mechanical Engineering, University Institute of Engineering and Technology, Maharshi Dayanand University, Rohtak, India
†Enzyme Technology and Protein Bioinformatics Laboratory, Department of Microbiology, Maharshi Dayanand University, Rohtak, India
**Corresponding Author: pratyoosh.shukla@gmail.com

1. INTRODUCTION

Neurodegenerative disorders are complex to understand and even more complicated to diagnose. Understanding patterns from the patient's history and training a machine to predict the diagnosis based on those patterns is projecting progressive practices to aid clinical procedures (Luo, 2016). Before elaborating on complicated artificial intelligence (AI)-based tools for neurodegenerative disorders such as Parkinson's disease, amyotrophic lateral sclerosis (ALS), Alzheimer's disease, epilepsy, seizures, strokes, and spinal cord injuries (SCIs), it is desirous to explain some important prerequisites of machine learning (ML) and wearables. Mitchell defines ML algorithms as "A computer program that is said to learn from experience E with respect to some class of tasks T and performance measure P, if its performance at tasks in T, as measured by P, improves with experience E" (Mitchell, 1997). In simple terms, ML is the state-of-the-art of learning and prediction from the dataset, where a dataset is defined as the pool of features. For neurodegenerative disorders the datasets are highly dynamic and unorganized.It is segmented in the form of electronic data from health records, imaging data from different modalities, data from high-throughput sequencing and data from various pathological and physiological analyses (Belle et al., 2015). In these situations, ML becomes interesting as it develops the ability to accomplish the task that underlies intelligence. Classification, regression, transcription, machine translation, anomaly detection, synthesis and sampling, denoising, density estimation and prediction are among such tasks for therapeutic of diseases and disorders (Goodfellow et al., 2016).

Furthermore, the field of AI has tremendously evolved since the introduction of sophisticated techniques and algorithms. Various AI-based algorithms were used for the purpose of ML can be differentiated between three categories: supervised learning,

---

ᵃ Both authors contributed equally and should be considered as first author.

Leveraging Biomedical and Healthcare Data
https://doi.org/10.1016/B978-0-12-809556-0.00005-8
77

unsupervised learning, and reinforced learning (Jordan and Mitchell, 2015). The core difference between these three techniques is the method of interpretation of the data. For supervised learning algorithms, the interpretations are done for the labeled featured data, whereas for unsupervised learning algorithms, the interpretation is based on exclusive learning and structure of data. On the other hand, the reinforced learning algorithm interacts with the data with a feedback loop between the learning systems and its experiences. To understand the mathematics behind these algorithms, readers can have a look at various available texts, as a detailed explanation on this platform is not a requisite (Nasrabadi, 2007; MacKay, 2003; Bradley, 1997). However, a table is provided that depicts the essential algorithms with their associated mechanisms and techniques for basic understanding (Table 1). Despite these pros, there also exist some challenges for these algorithms when operating with high dimensional and diverse scattered data, resulting in extensive computational cost. Therefore, deep learning platforms such as artificial neural networks (ANN) and convolutional neural networks (CNN) are designed to overcome the limitations associated with the formal ML techniques such as local constancy and smoothness regularization, dimensionality, and manifold learning.

For clinical assistance and disease diagnosis, the major challenge for ML is to create a model that not only predicts output based on the training data, but also on new inputs. Regularization is such a strategy that is used to explicitly train an ML algorithm to reduce the test error for a deep learning procedure (Goodfellow et al., 2016). Also, CNNs are remarkably successful for neural and medical applications as they use a distinct mathematical linear operator known as convolution. The application of deep learning requires knowledge of what an algorithm is and how it works. Also, a proper feedback mechanism is required to monitor and respond to the updated experimental procedure and clinical trials. Apart from that, the interpretation of volume, variety, and velocity of data, available at the various platforms, to form big data are equally important for an ML system to generate results from these algorithms in the form of speech recognition, computer vision, and language processing (Witten et al., 2016).

Optimization of data available at the various repositories is the next challenge for AI-based models, as the data presented in these repositories are very heterogeneous. Thus, a combination of traditional tools (such as operation research techniques and statistics), genetic algorithms (such as particle swarm optimization), and ML techniques (such as fuzzy logic) proves to be a convenient alternative (Zhu and Azar, 2015). Also, principal component analysis (PCA) is the distinctive method for analyzing this type of data due to its ability to handle bulky, noisy, and redundant variables. Using PCA, related variables can be converted to a set of uncorrelated variables (Abdi and Williams, 2010).

Table 1 Representation of essential ML algorithms with their associated working, mechanism, and application in neurology

S. no.	ML algorithm	Working mechanism	Application in clinical practices	References
1.	Regression	To find correlations among independent and dependent variables by curve fitting	Interpolation	Nasrabadi (2007)
2	Decision tree	To divide the populace among two or more homogeneous sets	Classification	Shaikhina et al. (2017)
3	Random forest	To classify a new object from attributes, each tree classifies from the collection of trees	Classification Prediction	Monteiro et al. (2018)
4	KNN	To predict via storing all available cases and classifies new cases through majority voting of k neighbors	Classification Interpolation	Zhang et al. (2016)
5	Logistic regression	To find the probability of occurrence of an event via logic function	Classification	Muhlestein et al. (2017)
6	K-means	To classify the dataset within a certain number of clusters	Clustering	Park et al. (2017)
7	Naïve Bayes	To predict by assuming the existence of an individual feature in a class that is dissimilar to the other feature	Classification	Bhagyashree et al. (2018)
8	SVM	To plot each data item as a point in n-dimensional space	Classification	Orru et al. (2012)
9	GBA	To predict via assembling of learning algorithms, which combines the prediction of several base estimators	Prediction	Eloyan et al. (2012)
10	DRA	To convert a set of data having vast dimensions into data with lesser dimensions	Classification Interpolation	Subasi and Gursoy (2010)

DBA, dimensionality reduction algorithm; *GBA*, gradient boosting algorithm; *KNN*, K-nearest neighbor; *SVM*, support vector machine.

2. WEARABLES AND ML-BASED THERAPEUTICS

Interpretation of biological data has become easier since the development of compact sensors (such as gyroscopes, surface electromyography sensors, accelerometers, magnetometers, electroencephalography sensors, and electrocardiography sensors). Point-of-care (POC) diagnosis at the patient's bedside to analyze these data and transmit it to the clinicians using cloud computing and ML is made easier with the advancements of wearables and smart phones (Kumari et al., 2017). Wearables are generally enfolded in tiny patches or in bandages or in any apparel able component. Pads, straps, rings, and wristbands are some of these health-monitoring devices that provide long-term uninterrupted recordings of electrophysiological action and acute physiological responses. They have extensively enhanced our perception of diseases, including PD and epilepsy. These flexible and stretchable electronic systems are powerful alternatives to bulky health-monitoring devices as they include light-emitting diodes, sensors, and different circuits that can interface with the skin and internal organs. Based on the mechanism of transduction, these devices can be classified as amperometric, potentiometric, calorimetric, piezoelectric, immunosensors, and optical. Ramasamy et al. have defined the basic architecture of the smart wireless wearable system with emphasize on aesthetics, water tolerance, power consumption, wireless communication, and the operating system (Ramasamy et al., 2014). Also, data processing using ML from these devices is done as per the prerequisite of a particular application.

The brain-machine interface that links the commands of the brain to the outside devices such as advanced computer programs aims to help handicapped and elderly people. As in neurodegenerative disorders, the brain and spinal cord are affected, but the ocular movements remain intact. Therefore, these movements can be used to generate commands for patient assistance using ML. Moreover, for patients with high-cervical SCI, it is desirous to reestablish the voluntary control of movements using an implantable cortex in the vertebrate for designing intracortical brain-computer interfaces. Ajiboye et al. provide the proof of concept of such an interface for nonhuman primates in which the Kalman filter, a denoising ML algorithm, is used to process the intracortical signals to decode the intentional movements of the patient (Ajiboye et al., 2017). Therefore, to design prosthetic limbs to treat patients with such neurodegenerative disorders, the integration of sensed biological data and ML techniques is solving the quest for alternative therapeutics.

3. NEURODEGENERATIVE THERAPEUTICS THROUGH AI

3.1 Parkinson's Disease

Parkinson's disease (PD) is a very common neurodegenerative disorder in which movements are affected due to genetic and environmental factors. Also, PD is difficult to diagnose in early stages, but due to ML, small vocals from a potential patient's voice recordings can be machine-interpreted (Gelb et al., 1999; Litvan et al., 1996). Using this information, it is easy to predict and diagnose PD. Therefore, the objective of AI-based clinical practices for PD is to achieve a differential diagnosis model for an individual. But these practices are more prone to errors as there exists a number of similar diseases (such as progressive supranuclear palsy, PSP) with similar symptoms (Tolosa et al., 2006). PSP patients resembles PD patients with some minor differences. The individual differential diagnoses of PD and PSP are done by means of a structural T-1 weighted MRI using a supervised ML method (Habeck et al., 2008; Salas-Gonzalez et al., 2010). This method includes image processing feature followed by feature extraction using PCA and classification algorithms. At last, classifiers are validated using cross-validation techniques such as k-fold cross validation. This method is used to evaluate analytical models by partitioning the original sample into a training set and a test set, where the training set is used to train the model and the test set is used to evaluate the model. The ML algorithm provides effectual discrimination of PD patients from PSP patients at an individual level, thus encouraging the application of computer-based diagnosis in clinical practices (Focke et al., 2011; Haller et al., 2013).

Moreover, PD is a chronic neurodegenerative disorder having symptoms such as tremors, rigidity, and postural instability, caused mainly by neuron disintegration of the substantia nigra (Hughes et al., 1992). However, essential tremor is a progressive neurological disorder of the CNS that occurs due to the effort exerted by muscles. In the last two decades, deep brain stimulation (DBS) has been recognized as an extremely effective remedy for the diagnosis of PD. DBS is a surgical process used to cure neurological disorders such as essential tremor even in later stages (Groiss et al., 2009). The present DBS system is an open-loop system with a decision-tree algorithm that uses a surgically implanted, battery-operated pulse generator to provide high-frequency electrical stimulation of the neurons that control movement. Also, in DBS, system stimulation is provided continuously with fixed parameters over time, which works on the basis of a feed-forward back-propagation neural network (NN) with overall high accuracy and high sensitivity for PD diagnosis via distinguishing movement and posture states based on trained classifiers (Shukla et al., 2014). As compared to the invasive surgical lessening methods, DBS has minimal tissue damage; because of this, it is a highly reversible process (Pilitsis et al., 2008; Kuncel and Grill, 2004). Currently, STN-DBS effectively treats the cardinal symptoms, which respond well to postural instability, rigidity, and tremor in the advanced stage. Also, for protecting PD patients from social isolation, DBS should be considered as a treatment that can be used at the initial stage.

Furthermore, wearables provide continuous measurement of key physiological parameters like such as movement of the body, along with data storage capacity and drug delivery, according to the command signal from the feedback therapy unit (Lotharius and Brundin, 2002). Multifunctional device like wearable-on-the-skin are available for diagnosis of patient with Parkinson disorders. These devices measures the movement disorders such as tremors using sensors (like silicon nano-membrane strain sensors) and then the monitored data are stored in different integrated memory chip. The blueprint of stored data will be analyzed and categorized into particular disease models, and then the analogous therapy will be provided through thermal stimuli at an optimized rate. Also, a temperature sensor is used to monitor the skin temperature simultaneously to prevent skin burning during drug delivery (Son et al., 2014). Despite these pros, wearable biomedical devices have some limitations within their own functioning, as they are not able to store the recorded data long term and can't deliver the advanced therapy for diagnostics.

3.2 Seizures and Epilepsy

Epilepsy is among the most common neurological disorders in which the human brain faces a number of transition states from normal interictal to preictal, then ictal and post-ictal (Lehnertz and Litt, 2005). Seizure prediction is a very complex task for diagnosis, but the introduction of ML techniques is used to achieve it in no mean time. Modern seizure prediction approaches enables feature extraction from intracranial state electroencephalographic (EEG) of the human brain and apply statistical analysis techniques for getting results (Mirowski et al., 2008). In the intracranial EEG, the local voltage potential is measured over brain cells using different electrodes, which is generated because of the different neuronal activities of the brain. These voltage signals have a lot of information about the mental defectiveness and neurological condition of the brain. A large amount of electrical discharge clearly indicates the signs of epilepsy (Kerem and Geva, 2017). Also, researchers advocate the use of bivariate features for seizure prediction, as these features measure the relationships between distant or neighboring EEG channels (Mirowski et al., 2008). A hybrid support vector machine (SVM) with particle swarm optimization (PSO) and a genetic algorithm (GA) is an efficient tool to detect epileptic seizures and classify the EEG data (Subasi et al., 2017). For better classification of data, the SVM design requires proper kernel parameters. The classified data are used to train a computer to gain knowledge of the nonlinear correlation among the features and their analogous labels. These trained computers test the EEG data quicker and more precisely and are also helpful in automating the system (Subasi, 2007). Moreover, the use of fuzzy logic with GA in the classification of epileptic signals is used to validate the remaining test datasets (Harikumar and Narayanan, 2003). Thus, systems based on PSO-SVM and GA-SVM are efficient in clinical practices as a real-time expert diagnostic tool.

3.3 Alzheimer's Disease

Alzheimer's disease (AD) is a brain disorder that starts mainly in middle life. The time-bound analysis of AD plays an important role in patient care, which is primarily associated with the detection of mild cognitive impairment (MCI) (Weiner et al., 2013). ML techniques are being used for advanced neuroimaging techniques such as positron emission tomography (PET) and magnetic resonance imaging (MRI), along with computational methods for better diagnosis (Liu et al., 2014). The deep CNN model is a fully trainable model that has the capability of capturing extremely nonlinear mappings among inputs and outputs. These models have computer vision and are inherently appropriate for image-related applications like image classification, segmentation, and denoising. Also a three-dimensional (3D) CNN model with a fully trained data modality with a nonlinear relationship is applied to predict the misplaced PET patterns (output) from the MRI data (input) (Li et al., 2014). The result obtained from this model outperformed the prior methods for disease diagnosis. Also, for volumetric image data interpretation, deep architecture models such as deep belief networks can be used instead of CNNs for better efficiency. The Alzheimer's disease Neuroimaging Initiative (ADNI) database is used for neuroimaging data, training, and testing of ML algorithms (Jack et al., 2008).

The designed architecture of deep learning techniques contains two primary components: stacked sparse autoencoders and a Softmax regression output layer that also assists the diagnosis of AD. The initial component sparse autoencoder consists of a neural network with a number of hidden layers to achieve deep representations of the original input vectors that are represented by neurons of each input layer (Le et al., 2011). In addition, the Softmax regression layer classifies instances by selecting the highest predicted probabilities of each label. Also, the Softmax layer uses a dissimilar start function from the other one, which is applied in the earlier hidden layers. This designed architecture has a multiobjective nature and could diminish the dependence on previous information in relation to the data (Bengio, 2009). Many ML methods have been projected for the treatment of AD using novel biomarkers such as β-amyloid (1–42) [Aβ (1–42)], total tau, and phospho-tau-181 (Humpel, 2011). Biomarkers act as an indicator, as the disease diagnostic utility of a biomarker is defined by its sensitivity and specificity (Salvatore et al., 2015). Thus, ML methods identify the patients with AD using them along with the risk prediction of MCI (Andreasen and Blennow, 2005).

3.4 Amyotrophic Lateral Sclerosis

Amyotrophic lateral sclerosis (ALS) is a neurodegenerative disease with uncertain pathogenesis that is characterized by progressive muscle weakness. It is also known as motor neuron disease. Two important issues related to ALS are the estimation of the cognitive status and the restoration of communication with the external environment (Geronimo et al., 2016). The human brain is filled with neurons and nerve cells, which are interconnected by axons and dendrites. Every activity that a human performs needs neurons to work, and is carried out in the form of small electrical signals. In the course of diagnosis, there exists an effective communication tool for ALS-affected patients that is known as brain-computer interface (BCI) technology (Riccio et al., 2013). BCI works in the same way as our brain by decoding the brain signals into commands for the purpose of movement restoration using advanced ML algorithms for computer interpretation. In the BCI technique, we implant a set of electrodes into the gray matter of the brain itself and then these electrodes read the brain signals by measuring the voltage difference between the neurons. This allows ALS patients to communicate with the social environment without any contribution of the muscles and peripheral nerves (Marchetti and Priftis, 2015). Both BCI systems, invasive (such as electroencephalographic, sensor motor rhythms, and slow cortical potentials) and noninvasive (such as magnetoencephalography and functional magnetoresonance imaging) provide a possible means for communication for disabled persons via decoding signals. An alternative BCI model based on motor imagery that uses the modulation of oscillations produced in the motor cortex helps the person with ALS for communication purposes (Schomer and Da Silva, 2012). Apart from it, there also exists a deep learning framework ALS.AI, which is designed to generate and apply algorithms to pinpoint possible causes of the disease and predict patient responses to candidate drugs. As the present, the available methods of diagnosis are limited, suggesting an urgent need to develop new treatments for ALS using such deep learning protocols.

3.5 Stroke and Spinal Cord Injury

A stroke is the interruption of blood supply to a part of the brain or the rupture of a blood vessel in the brain, causing hemorrhage. Treatments for both these types of strokes are time sensitive (Hacke et al., 2004). However, patients commonly do not recognize the symptoms, and as a result, these treatments are largely underutilized (Prabhakaran et al., 2012). Using AI, an individual's neurological functions such as movement abnormalities are continuously monitored; the presence of any alarming signal activates the emergency medical services, even without the patient's knowledge (Acampora et al., 2013). Ambient intelligence (that is, the combination of AI and pervasive healthcare) in conjunction with telestroke networks (Müller-Barna et al., 2012), automated imaging interpretation, and a prehospital thrombolytic, can manage the acute stroke by reducing the delay period. ANN is a mathematical and computational model that uses a connection's computational approach for information processing and prediction (Cheng et al., 2014). It utilizes the patient's parameters such as age, sex, and stroke severity along with the treatment time for prediction. For a stroke, it is used to predict thrombolysis and surgical outcomes for the patient and his family. The time-bound thrombolysis treatment also leads to a better recovery with an increased survival rate (Walter et al., 2012). In a study conducted on a set of patients who had strokes, it was found that the sensitivity and specificity of ANN models can be used to construct a receiver operating characteristics (ROC) curve (where the data interpolation is made using logistic regression) for better prediction of outcomes.

Also, patients with SCIs have severe motor disabilities in the lower as well as the upper limbs, and such people rely on the help of others in everyday life. Restoration of the grasp function can help them to gain some independence. BCI is such a platform available for this purpose. However, noninvasive EEG-based neuroprosthetics is another way to aid quadriplegics in restoring different grasping movements (Müller-Putz et al., 2017). In this discourse, the EEG is recorded via signal processing methods and ML principles transfer them into control signals to restore, replace, and improve the functioning of the CNS.

4. AI-BASED CLINICAL DECISION-MAKING

Contemporary healthcare practices, including the complex algorithms of ML, also provide a platform for affordable AI-based clinical decision systems (CDSs) along with POC diagnosis. AI techniques such as fuzzy logic, fuzzy-genetic algorithm (fuzzy-GA), and rule-based reasoning build CDSs that focus on computer-oriented therapeutics. Also, a growing myriad of treatment options and tools such as deep learning facilitates a new generation of applications focused on precision medicine (Vashistha et al., 2018). In this direction, Bennett et al. provides an AI simulation-based framework that optimizes decisions in complicated and ambiguous situations for personalized medicine. In their study, they combined Markov decision processes (MDP) and dynamic decision networks for developing complex plots by simulations from electronic health records of the patients. Also, the cost per unit of outcome was minimized substantially from $497 to $189, the study concluded (Bennett and Hauser, 2013). Thus, the use of AI for CDS is conclusive to argue further, especially for neurological disorders where the discourse is even more tedious and expensive.

Furthermore, the availability of a pool of data in the form of public repositories enables analysis of the effectiveness of the researched methodology and its validation for CDS. One such study that included AI-based CDS employs a combinatorial intelligent system for the prediction of PD development by ML. Here, expectation maximization algorithms and PCA are engaged to address the clustering of data and multicollinearity problems in the experimental datasets, followed by the use of software tools [such as support vector regression (SVR) and the adaptive neuro-fuzzy inference system (ANFIS)] for prediction of PD advancement. Also, when used in combination, genetic data (such as single nucleotide polymorphism) and multimodality neuroimaging (such as MRI and PET) provide insights for AD risk factors. As this data integration is heterogeneous, encompassing an effective model using deep feature learning and a fusion framework can be used for CDS (Zhou et al., 2017). Another method to probe medical information from the EHR for prediction and progression of motor neuron diseases, supporting AI-based CDS, uses Hbase and the random forest classifier for ALS diagnosis. Consequently, this platform provides a solution to the challenges of next-generation therapeutics, which will be in the scope of every section of society, as shown in Fig. 1.

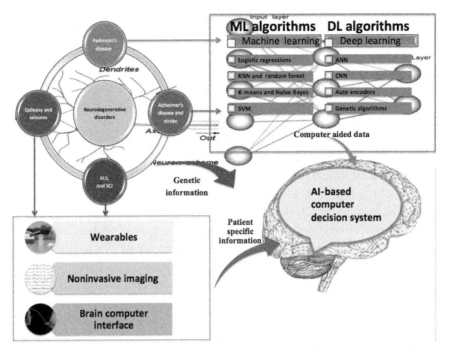

Fig. 1 Schematic representation for AI-based clinical decision system for next generation therapeutics.

5. LIMITATIONS AND FUTURE PERSPECTIVES

Unconventional AI algorithms have the potential to diagnose acute complex neurode-generative emergencies within the optimized time framework. Also, chronic disorders can be recognized early, and care may be individualized along with improved quality of care and a significantly reduced cost. However, there also exist some limitations to make this platform a generalized one, including a lack of international biomedical information sharing network platforms, fewer specialists with both clinical and programming expertise, a lack of reliable standards for communication and data interchange, a conventional education syllabus of medical specialists, lack of training of ML algorithms from a diverse set of international data, and morals (such as legal, privacy, ethical) confrontation. Also, international bodies need to create profile-raising policies to fast-track investment in AI by rewarding the hospitals and physicians. Overcoming these limitations can eventually pay off the initial monetary investments to create a path for employing AI in real world medicine.

REFERENCES

Abdi, H., Williams, L.J., 2010. Principal component analysis. Wiley Interdiscip. Rev. Comput. Stat. 2, 433–459.

Acampora, G., Cook, D.J., Rashidi, P., Vasilakos, A.V., 2013. A survey on ambient intelligence in health-care. Proc. IEEE 101, 2470–2494.

Ajiboye, A.B., Willett, F.R., Young, D.R., Memberg, W.D., Murphy, B.A., Miller, J.P., Walter, B.L., Sweet, J.A., Hoyen, H.A., Keith, M.W., Peckham, P.H., 2017. Restoration of reaching and grasping movements through brain-controlled muscle stimulation in a person with tetraplegia: a proof-of-concept demonstration. Lancet 389, 1821–1830.

Andreasen, N., Blennow, K., 2005. CSF biomarkers for mild cognitive impairment and early Alzheimer's disease. Clin. Neurol. Neurosurg. 107, 165–173.

Belle, A., Thiagarajan, R., Soroushmehr, S.M., Navidi, F., Beard, D.A., Najarian, K., 2015. Big data analytics in healthcare. BioMed Res. Int. 370194

Bengio, Y., 2009. Learning Deep Architectures for AI. vol. 2. Foundations and Trends® in Machine Learning. Now Publishers Inc., Boston - Delft, pp. 1–127

Bennett, C.C., Hauser, K., 2013. Artificial intelligence framework for simulating clinical decision-making: a Markov decision process approach. Artif. Intell. Med. 57, 9–19.

Bhagyashree, S.I.R., Nagaraj, K., Prince, M., Fall, C.H., Krishna, M., 2018. Diagnosis of dementia by machine learning methods in epidemiological studies: a pilot exploratory study from South India. Soc. Psychiatry Psychiatr. Epidemiol. 53, 77–86.

Bradley, A.P., 1997. The use of the area under the ROC curve in the evaluation of machine learning algorithms. Pattern Recogn. 30, 1145–1159.

Cheng, C.A., Lin, Y.C., Chiu, H.W., 2014. Prediction of the prognosis of ischemic stroke patients after intravenous thrombolysis using artificial neural networks, In: Mantas, J., Househ, M.S., Hasman, A. (Eds.), Integrating Information Technology and Management for Quality of Care, pp. 115–118.

Eloyan, A., Muschelli, J., Nebel, M.B., Liu, H., Han, F., Zhao, T., Barber, A.D., Joel, S., Pekar, J.J., Mostofsky, S.H., Caffo, B., 2012. Automated diagnoses of attention deficit hyperactive disorder using magnetic resonance imaging. Front. Syst. Neurosci. 6, 61.

Focke, N.K., Helms, G., Scheewe, S., Pantel, P.M., Bachmann, C.G., Dechent, P., Ebentheuer, J., Mohr, A., Paulus, W., Trenkwalder, C., 2011. Individual voxel based subtype prediction can differentiate progressive supranuclear palsy from idiopathic Parkinson syndrome and healthy controls. Hum. Brain Mapp. 32, 1905–1915.

Gelb, D.J., Oliver, E., Gilman, S., 1999. Diagnostic criteria for Parkinson disease. Arch. Neurol. 56, 33–39.

Geronimo, A., Simmons, Z., Schiff, S.J., 2016. Performance predictors of brain–computer interfaces in patients with amyotrophic lateral sclerosis. J. Neural Eng. 13, 026002.

Goodfellow, I., Bengio, Y., Courville, A., Bengio, Y., 2016. Deep Learning. vol. 1. MIT Press, Cambridge.

Groiss, S.J., Wojtecki, L., Südmeyer, M., Schnitzler, A., 2009. Deep brain stimulation in Parkinson's disease. Ther. Adv. Neurol. Disord. 2, 379–391.

Habeck, C., Foster, N.L., Perneczky, R., Kurz, A., Alexopoulos, P., Koeppe, R.A., Drzezga, A., Stern, Y., 2008. Multivariate and univariate neuroimaging biomarkers of Alzheimer's disease. NeuroImage 40, 1503–1515.

Hacke, W., Donnan, G., Fieschi, C., Kaste, M., Broderick, J.P., Brott, T., Frankel, M., Grotta, J.C., Haley, J.E., Kwiatkowski, T., Levine, S.R., 2004. Association of outcome with early stroke treatment: pooled analysis of ATLANTIS, ECASS, and NINDS rt-PA stroke trials. Lancet 9411, 768–774.

Haller, S., Badoud, S., Nguyen, D., Barnaure, I., Montandon, M.L., Lovblad, K.O., Burkhard, P.R., 2013. Differentiation between Parkinson disease and other forms of Parkinsonism using support vector machine analysis of susceptibility-weighted imaging (SWI): initial results. Eur. Radiol. 23, 12–19.

Harikumar, R., Narayanan, B.S., 2003. In: Fuzzy techniques for classification of epilepsy risk level from EEG signals.Conference on Convergent Technologies for the Asia-Pacific Region. vol. 1, pp. 209–213.

Hughes, A.J., Ben-Shlomo, Y., Daniel, S.E., Lees, A.J., 1992. What features improve the accuracy of clinical diagnosis in Parkinson's disease: a clinicopathologic study. Neurology 42, 1142.

Humpel, C., 2011. Identifying and validating biomarkers for Alzheimer's disease. Trends Biotechnol. 29, 26–32.

Jack, C.R., Bernstein, M.A., Fox, N.C., Thompson, P., Alexander, G., Harvey, D., Borowski, B., Britson, P.J., Whitwell, J.L., Ward, C., Dale, A.M., 2008. The Alzheimer's disease neuroimaging initiative (ADNI): MRI methods. J. Magn. Reson. Imaging 27, 685–691.

Jordan, M.I., Mitchell, T.M., 2015. Machine learning: trends, perspectives, and prospects. Science 349, 255–260.

Kerem, D.H., Geva, A.B., 2017. Brain state identification and forecasting of acute pathology using unsupervised fuzzy clustering of eeg temporal patterns. In: Fuzzy and Neuro-Fuzzy Systems in Medicine. CRC Press, pp. 19–68.

Kumari, P., Mathew, L., Syal, P., 2017. Increasing trend of wearables and multimodal interface for human activity monitoring: a review. Biosens. Bioelectron. 90, 298–307.

Kuncel, A.M., Grill, W.M., 2004. Selection of stimulus parameters for deep brain stimulation. Clin. Neurophysiol. 115, 2431–2441.

Le, Q.V., Ngiam, J., Coates, A., Lahiri, A., Prochnow, B., Ng, A.Y., 2011. On optimization methods for deep learning. In: Proceedings of the 28th International Conference on International Conference on Machine Learning. Omnipress, pp. 265–272.

Lehnertz, K., Litt, B., 2005. The first international collaborative workshop on seizure prediction: summary and data description. Clin. Neurophysiol. 116, 493–505.

Li, R., Zhang, W., Suk, H.I., Wang, L., Li, J., Shen, D., Ji, S., 2014. Deep learning based imaging data completion for improved brain disease diagnosis. In: International Conference on Medical Image Computing and Computer-Assisted Intervention. Springer, Cham, pp. 305–312.

Litvan, I., Agid, Y., Calne, D., Campbell, G., Dubois, B., Duvoisin, R.C., Goetz, C.G., Golbe, L.I., Grafman, J., Growdon, J.H., Hallett, M., 1996. Clinical research criteria for the diagnosis of progressive supranuclear palsy (Steele-Richardson-Olszewski syndrome) report of the NINDS-SPSP international workshop. Neurology 47, 1–9.

Liu, S., Liu, S., Cai, W., Pujol, S., Kikinis, R., Feng, D., 2014. Early diagnosis of Alzheimer's disease with deep learning. In: IEEE 11th International Symposium on Biomedical Imaging, pp. 1015–1018.

Lotharius, J., Brundin, P., 2002. Pathogenesis of Parkinson's disease: dopamine, vesicles and α-synuclein. Nat. Rev. Neurosci. 3, 932.

Luo, G., 2016. PredicT-ML: a tool for automating machine learning model building with big clinical data. Health Inf. Sci. Syst. 4, 5.

MacKay, D.J., 2003. Information Theory, Inference and Learning Algorithms. Cambridge university Press, Cambridge.

Marchetti, M., Priftis, K., 2015. Brain–computer interfaces in amyotrophic lateral sclerosis: a metanalysis. Clin. Neurophysiol. 126, 1255–1263.

Mirowski, P.W., LeCun, Y., Madhavan, D., Kuzniecky, R., 2008. In: Comparing SVM and convolutional networks for epileptic seizure prediction from intracranial EEG.IEEE Workshop on Machine Learning for Signal Processing, pp. 244–249.

Mitchell, T.M., 1997. Machine Learning. McGraw-Hill International Editions Computer Science Series.

Monteiro, M.A.B., Fonseca, A.C., Freitas, A.T., e Melo, T.P., Francisco, A.P., Ferro, J.M., Oliveira, A., 2018. Using machine learning to improve the prediction of functional outcome in ischemic stroke patients. IEEE/ACM Trans. Comput. Biol. Bioinf. 1–1.

Muhlestein, W.E., Morone, P.J., Kallos, J.A., Chambless, L.B., 2017. Using logistic regression and a novel machine learning technique to predict discharge status after craniotomy for meningioma. J. Neurol. Surg. B: Skull Base 78, A034.

Müller-Barna, P., Schwamm, L.H., Haberl, R.L., 2012. Telestroke increases use of acute stroke therapy. Curr. Opin. Neurol. 25, 5–10.

Müller-Putz, G.R., Plank, P., Stadlbauer, B., Statthaler, K., Uroko, J.B., 2017. 15 Years of evolution of non-invasive EEG-based methods for restoring hand & arm function with motor neuroprosthetics in individuals with high spinal cord injury: a review of Graz BCI research. J. Biomed. Sci. Eng. 10, 317.

Nasrabadi, N.M., 2007. Pattern recognition and machine learning. J. Electron. Imaging. 16, 049901.

Orru, G., Pettersson-Yeo, W., Marquand, A.F., Sartori, G., Mechelli, A., 2012. Using support vector machine to identify imaging biomarkers of neurological and psychiatric disease: a critical review. Neurosci. Biobehav. Rev. 36, 1140–1152.

Park, E., Chang, H.J., Nam, H.S., 2017. Use of machine learning classifiers and sensor data to detect neurological deficit in stroke patients. J. Med. Internet Res. 19, 1–4

Pilitsis, J.G., Chu, Y., Kordower, J., Bergen, D.C., Cochran, E.J., Bakay, R.A., 2008. Postmortem study of deep brain stimulation of the anterior thalamus: case report. Neurosurgery 62, E530–E532.

Prabhakaran, S., McNulty, M., O'neill, K., Ouyang, B., 2012. Intravenous thrombolysis for stroke increases over time at primary stroke centers. Stroke 43, 875–877.

Ramasamy, V., Gowda, C., Noopuran, S., 2014. The Basics of Designing Wearable Electronics With Microcontrollers. Cypress Semiconductor, p. 17.

Riccio, A., Simione, L., Schettini, F., Pizzimenti, A., Inghilleri, M., Olivetti Belardinelli, M., Mattia, D., Cincotti, F., 2013. Attention and P300-based BCI performance in people with amyotrophic lateral sclerosis. Front. Hum. Neurosci. 7, 732.

Salas-Gonzalez, D., Gorriz, J.M., Ramirez, J., Illán, I.A., López, M., Segovia, F., Chaves, R., Padilla, P., Puntonet, C.G., Initiative, A.'s.D.N., 2010. Feature selection using factor analysis for Alzheimer's diagnosis using PET images. Med. Phys. 37, 6084–6095.

Salvatore, C., Cerasa, A., Battista, P., Gilardi, M.C., Quattrone, A., Castiglioni, I., 2015. Magnetic resonance imaging biomarkers for the early diagnosis of Alzheimer's disease: a machine learning approach. Front. Neurosci. 9, 307.

Schomer, D.L., Da Silva, F.L., 2012. Niedermeyer's Electroencephalography: Basic Principles, Clinical Applications, and Related Fields. Lippincott Williams & Wilkins, Philadelphia.

Shaikhina, T., Lowe, D., Daga, S., Briggs, D., Higgins, R., Khovanova, N., 2017. Decision tree and random forest models for outcome prediction in antibody incompatible kidney transplantation. Biomed. Signal Process. Control. https://doi.org/10.1016/j.bspc.2017.01.012.

Shukla, P., Basu, I., Tuninetti, D., 2014. In: Towards closed-loop deep brain stimulation: decision tree-based essential tremor patient's state classifier and tremor reappearance predictor.36th Annual International Conference of the IEEE Engineering in Medicine and Biology Society, pp. 2605–2608.

Son, D., Lee, J., Qiao, S., Ghaffari, R., Kim, J., Lee, J.E., Song, C., Kim, S.J., Lee, D.J., Jun, S.W., Yang, S., 2014. Multifunctional wearable devices for diagnosis and therapy of movement disorders. Nat. Nanotechnol. 9, 397.

Subasi, A., 2007. EEG signal classification using wavelet feature extraction and a mixture of expert model. Expert Syst. Appl. 32, 1084–1093.

Subasi, A., Gursoy, M.I., 2010. EEG signal classification using PCA, ICA, LDA and support vector machines. Expert Syst. Appl. 37, 8659–8666.

Subasi, A., Kevric, J., Canbaz, M.A., 2017. Epileptic seizure detection using hybrid machine learning methods. Neural Comput. Appl, 1–9.

Tolosa, E., Wenning, G., Poewe, W., 2006. The diagnosis of Parkinson's disease. Lancet Neurol. 5, 75–86.

Vashistha, R., Chhabra, D., Shukla, P., 2018. Integrated artificial intelligence approaches for disease diagnostics. Indian J. Microbiol, 1–4.

Walter, S., Kostopoulos, P., Haass, A., Keller, I., Lesmeister, M., Schlechtriemen, T., Roth, C., Papanagiotou, P., Grunwald, I., Schumacher, H., Helwig, S., 2012. Diagnosis and treatment of patients with stroke in a mobile stroke unit versus in hospital: a randomised controlled trial. Lancet Neurol. 11, 397–404.

Weiner, M.W., Veitch, D.P., Aisen, P.S., Beckett, L.A., Cairns, N.J., Green, R.C., Harvey, D., Jack, C.R., Jagust, W., Liu, E., Morris, J.C., 2013. The Alzheimer's Disease Neuroimaging Initiative: a review of papers published since its inception. Alzheimer's Dement. 9, 111–e194.

Witten, I.H., Frank, E., Hall, M.A., Pal, C.J., 2016. Data Mining: Practical Machine Learning Tools and Techniques. Morgan Kaufmann, USA.

Zhang, Y., Lu, S., Zhou, X., Yang, M., Wu, L., Liu, B., Phillips, P., Wang, S., 2016. Comparison of machine learning methods for stationary wavelet entropy-based multiple sclerosis detection: decision tree, k-nearest neighbors, and support vector machine. SIMULATION 92, 861–871.

Zhou, T., Thung, K.H., Zhu, X., Shen, D., 2017. In: Feature learning and fusion of multimodality neuroimaging and genetic data for multi-status dementia diagnosis. International Workshop on Machine Learning in Medical Imagin. Springer, Cham, pp. 132–140.

Zhu, Q., Azar, A.T.e., 2015. Complex System Modelling and Control through Intelligent Soft Computations. Springer, Germany.

CHAPTER 6

Robust Detection of Epilepsy Using Weighted-Permutation Entropy: Methods and Analysis

Bilal Fadlallah*, Ali Fadlallah[†], Mahdi Razafsha[‡], Nabil Karnib[§], Kevin Wang[¶], Firas Kobeissy[§,¶]

[*]Department of Biomedical Engineering at Georgia Tech and Emory, Georgia Institute of Technology, Atlanta, GA, United States
[†]Northern General Hospital, Sheffield, United Kingdom
[‡]Department of Psychiatry, McKnight Brain Institute, University of Florida, Gainesville, FL, United States
[§]Department of Biochemistry and Molecular Genetics, American University of Beirut, Beirut, Lebanon
[¶]Program for Neurotrauma, Neuroproteomics & Biomarkers Research, Department of Emergency Medicine, University of Florida, Gainesville, FL, United States

1. INTRODUCTION

1.1 Background

Epilepsy, defined as recurrent epileptic seizures that are unprovoked by any immediate identified cause (Commission on Epidemiology and Prognosis, International League Against Epilepsy, 1993), affects up to 1% of the population worldwide and comes second to stroke as a serious neurological disorder impacting the lives of millions (Stafstrom, 2006; World Health Organization, 2012). There are an estimated 65 million epilepsy patients in the world today, with >2 million Americans having the disorder. Many of these patients suffer from seizures that cannot be controlled by the currently available antiepileptic medications. The total burden due to epilepsy is around 0.5% of the global burden of disease with approximately 1% of the total days lost due to disease. The signs and symptoms of epilepsy vary due to the involvement of several areas of the cortex and underlying brain systems (Lytton, 2008), and one of the most devastating features of epilepsy is the apparently unpredictable nature of the seizures (Litt and Echauz, 2002). However, the recent literature supports the notion that seizures begin before the onset of clinical signs and symptoms; thus, its early and accurate detection is important to improve the quality of life of epileptic sufferers (Litt and Echauz, 2002; McSharry et al., 2003; Minasyan et al., 2010). Therefore, the real challenge is in the development of highly reliable methods for seizure onset detection prior to clinical syndromes. Amplitude-

---

☆ This chapter is an edited reprint of the paper by Fadlallah et al. (2013) Physical Review E 87, 2, 022911, titled: "Weighted-permutation entropy: A complexity measure for time series incorporating amplitude information."

Leveraging Biomedical and Healthcare Data
https://doi.org/10.1016/B978-0-12-809556-0.00006-X

integrated electroencephalography (aEEG) is one technique that is useful to monitor brain activity; however, the distinction of seizure activity of EEG requires a specialist and may not have high specificity and sensitivity (Shah et al., 2008; Hellström-Westas et al., 2006). As the difficulty in interpreting aEEG in clinical practice relates mainly to the distinction between artifact and activity (Hagmann et al., 2006), several mathematical algorithms were incorporated for its analysis (Tao and Mathur, 2010). The current chapter will emphasize developing algorithmic methods for analyzing epileptic data and extracting specific quantifiers that can be used for its prediction.

1.2 Approach

Automatically detecting epilepsy can be problematic, especially in the absence of seizures. Several factors, including EEG artifacts, noise sources, and muscle movements, might also complicate signal analysis. Most of the currently used methods consist of two stages: a feature extraction stage and a classification stage, which can be categorized as time domain, frequency domain, or time-frequency domain methods. In this chapter, we propose to use measures of signal complexity and regularity for epilepsy detection. We adopt a sliding window approach and investigate the effect of the free parameters involved. Signal complexity can be easily captured using entropy measures. Entropy is an information theoretic measure that captures the variability of a signal. The recently introduced concept of permutation entropy or PE (Bandt and Pompe, 2002) allows estimating the complexity of a time series using ordinal descriptors. Weighted-permutation entropy or WPE (Fadlallah et al., 2013) extends this concept to incorporate amplitude information from time series and is hence suitable for detecting complexity changes in time series. The method is compared to several well-known algorithms in the literature, namely approximate entropy or AE (Pincus, 1991) and the composite PE index or CPEI (Olofsen et al., 2008). The data used in our experiments includes (but is not limited to) the one used by Quiroga et al. (2002).

2. COMPUTATIONAL DETAILS

2.1 Motivation

There is little consensus on the definition of a signal's complexity. Among the different approaches, entropy-based ones are inspired by either nonlinear dynamic (Pincus, 1991) or symbolic dynamics (Bandt and Pompe, 2002; Kurths et al., 1996). Permutation entropy (PE) has been recently suggested as a complexity measure based on comparing neighboring values of each point and mapping them to ordinal patterns (Bandt and Pompe, 2002). Using ordinal descriptors is helpful in the sense that it adds immunity to large artifacts occurring with low frequencies. The usage of permutation entropy has spanned regular, chaotic, noisy, or real-world time series; specifically, it has been employed in the context of neural (Li et al., 2011), electroencephalographic (Li et al., 2007, 2008; Bruzzo et al., 2008; Cao et al., 2004), electrocardiographic (Daoming et al., 2008; Graff et al., 2012), and stock market time series (Zunino et al., 2009). In this chapter, we suggest a modification that alters the way PE handles the patterns extracted from a given signal by incorporating amplitude information. For many time series of interest, the new scheme better tracks abrupt changes in the signal and assigns less complexity to segments that exhibit regularity or are subject to noise effects. Examples include any time series containing amplitude coded information. For such signals, the suggested method has the advantage of providing immunity to degradation by noise and (linear) distortion.

The chapter is organized as follows. In Sections 2.2 and 2.3, we briefly introduce permutation entropy and formulate weighted-permutation entropy. Experimental details are presented in Section 3 on synthetic, single channel, and dense-array EEG, and finally epileptic data. Section 4 offers discussion and concluding remarks.

2.2 Permutation Entropy

Consider the time series $\{x_t\}_{t=1}^T$ and its time-delay embedding representation $X_j^{m,\tau} = \{x_j, x_{j+\tau}, \ldots, x_{j+(m-1)\tau}\}$ for $j = 1, 2, \ldots, T - (m-1)\tau$ where m and τ denote, respectively, the embedding dimension and time delay. To compute PE, each of the $N = T - (m-1)\tau$ subvectors is assigned a single motif out of $m!$ possible ones (representing all unique orderings of m different real numbers). PE is then defined as the Shannon entropy of the $m!$ distinct symbols $\{\pi_i^{m,\tau}\}_{i=1}^{m!}$ denoted as Π:

$$H(m, \tau) = -\sum_{i:\pi_i^{m,\tau} \in \Pi} p(\pi_i^{m,\tau}) \ln p(\pi_i^{m,\tau}) \tag{1}$$

$p(\pi_i^{m,\tau})$ is defined as:

$$p(\pi_i^{m,\tau}) = \frac{\left\| \left\{ j : j \leq N, \text{type}\left(X_j^{m,\tau} \right) = \pi_i^{m,\tau} \right\} \right\|}{N} \tag{2}$$

where type $(\cdot)$ denotes the map from pattern space to symbol space and $\|\cdot\|$ denotes the cardinality of a set. An alternative way of writing $p(\pi_i^{m,\tau})$ is:

$$p(\pi_i^{m,\tau}) = \frac{\sum_{j \leq N} 1_{u:\text{type}(u)=\pi_i}\left(X_j^{m,\tau} \right)}{\sum_{j \leq N} 1_{u:\text{type}(u)\in\Pi}\left(X_j^{m,\tau} \right)} \tag{3}$$

where $1_A(u)$ denotes the indicator function of set A defined as $1_A(u) = 1$ if $u \in A$ and $1_A(u) = 0$ if $u \notin A$. PE assumes values in the range $[0, \ln m!]$ and is invariant under nonlinear monotonic transformations.

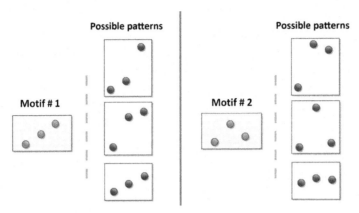

Fig. 1 Two examples of possible *m*-dimensional vectors corresponding to the same motif. The value of *m* used is 3.

The main shortcoming in the above definition of PE resides in the fact that no information besides the order structure is retained when extracting the ordinal patterns for each time series. This may be inconvenient for the following reasons: (i) most time series have information in the amplitude that might be lost when solely extracting the ordinal structure, (ii) ordinal patterns where the amplitude differences between the time series points are greater than others should not contribute similarly to the final PE value, and (iii) ordinal patterns resulting from small fluctuations in the time series can be due to the effect of noise and should not be weighted uniformly toward the final value of PE. Fig. 1 shows how the same ordinal pattern can originate from different *m*-dimensional vectors.

2.3 Weighted-Permutation Entropy

To counterweight these facts, we propose a modification of the current PE procedure to incorporate significant information from the time series when retrieving the ordinal patterns. The main motivation lies in saving useful amplitude information carried by the signal. We refer to this procedure as weighted–permutation entropy (WPE) and summarize it in the following steps. First, the weighted relative frequencies for each motif are calculated as follows:

$$p_w\left(\pi_i^{m,\tau}\right) = \frac{\sum_{j \leq N} 1_{u:\text{type}(u)=\pi_i}\left(X_j^{m,\tau}\right) \cdot w_j}{\sum_{j \leq N} 1_{u:\text{type}(u)\in\Pi}\left(X_j^{m,\tau}\right) \cdot w_j} \tag{4}$$

WPE is then computed as:

$$H_w(m,\tau) = -\sum_{i:\pi_i^{m,\tau}\in\Pi} p_w\left(\pi_i^{m,\tau}\right) \ln p_w\left(\pi_i^{m,\tau}\right) \tag{5}$$

Note that when $w_j = \beta \ \forall j \leq N$ and $\beta > 0$, WPE reduces to PE. It is also interesting to highlight the difference between the definition of weighted entropy in this context and previous ones suggested in the literature. Weighted entropy, defined as $H_{we} = -\sum_k w_k p_k \ln p_k$, has been suggested as a variant to entropy that uses a probabilistic experiment whose elementary events are characterized by weights w_k (Guiasu, 1971). WPE, on the other hand, extends the concept of PE while keeping the same Shannon's entropy expression reflected by Litt and Echauz (2002), hence weights are added prior to computing the $p(\pi_i^{m,\tau})$. The choice of weight values Wi is equivalent to selecting a specific (or combination of) feature(s) from each vector $X_j^{m,\tau}$. Such features may differ according to the context used. Note that the relation $\sum_i p_w(\pi_i^{m,\tau}) = 1$ still holds. In this chapter, we use the variance or energy of each neighbor's vector $X_j^{m,\tau}$ to compute the weights. Let $\overline{X}_j^{m,\tau}$ denote the arithmetic mean of $X_j^{m,\tau}$ or:

$$\overline{X}_j^{m,\tau} = \frac{1}{m} \sum_{k=1}^{m} x_{j+(k+1)\tau} \tag{6}$$

We can hence express each weight value as:

$$w_j = \frac{1}{m} \sum_{k=1}^{m} \left(x_{j+(k-1)\tau} - \overline{X}_j^{m,\tau} \right)^2 \tag{7}$$

The motivation behind this setting is to specifically counteract the limitations discussed in the previous section, that is, weight differently neighboring vectors having the same ordinal patterns but different amplitude variations. In this way, WPE can be also used to detect abrupt changes in noisy or multicomponent signals. The modified $p(\pi_i^{m,\tau})$ can be then thought of as the proportion of variance accounted for by each motif. The above definition of WPE retains most of PE's properties and is invariant under affine linear transformations. WPE, however, presents a specificity, given that it incorporates amplitude information and demonstrates more robustness to noise.

3. EXPERIMENTS AND RESULTS

3.1 Synthetic Data

As a first motivation, we suggest analyzing the behavior of PE and WPE in the presence of an impulsive and noisy signal. Fig. 2 shows 1000 samples of a signal consisting of an impulse and additive white Gaussian noise (AWGN) with zero mean and unit variance. Windows of 80 samples slid by 10 samples were used and the results were averaged over 10 simulations. A remarkable drop in the value of WPE is noticed in the impulse region. No marked change can be observed in the case of PE for the same region. As a next step, we try a train of Gaussian-modulated sinusoidal pulses with decaying amplitudes. The value of τ was set to 1. Sliding windows of 50 samples with increments of 10 samples

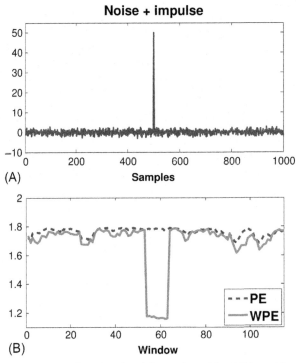

Fig. 2 PE versus WPE in the case of an impulse. (A) Impulse with additive white Gaussian noise with zero mean and unit variance. (B) Computed PE and WPE values with windows of 80 samples slid by 10 samples. A remarkable drop in the values of WPE is noticed in the impulse region for which PE values do not show any marked change.

were used and m was set to 3. Again, the signal was corrupted by AWGN and simulations were run across different variance levels.

Fig. 3 shows the variations of the signal's entropy for four different methods. The performance of PE and WPE is compared to two other methods from the literature, namely approximate entropy or ApEn (Pincus, 1991; Chon et al., 2009) and the composite PE index or CPEI (Olofsen et al., 2008). In the following, we give a brief description of each. Approximate entropy (ApEn or AE) is a measure that quantifies the regularity or predictability of a time series. It is defined with respect to a free parameter r as follows:

$$H_a = \Phi^m(r) - \Phi^{m+1}(r) \tag{8}$$

where $\Phi^m(r)$ is defined as:

$$\Phi^m(r) = \frac{1}{N-(m-1)\tau} \sum_{i=1}^{N-(m-1)\tau} \ln C_i^m(r) \tag{9}$$

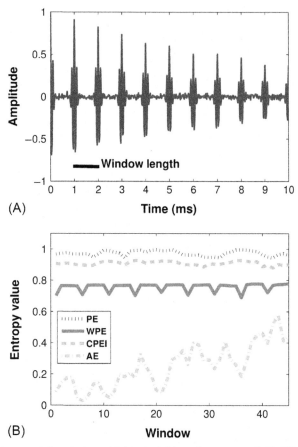

Fig. 3 (A) A Gaussian-modulated sinusoidal train with a frequency of 10 kHz, a pulse repetition frequency of 1 kHz, and an amplitude attenuation rate of 0.9. Initial signal was corrupted by additive white Gaussian noise (AWGN) having mean $\mu = 0$ and variance $\sigma^2 = 0.2$. The sampling rate was 50 kHz and computations used a 50-sample sliding window with increments of 10 samples. The recorded SNR was of 4.8 dB. (B) Different entropy measures (PE, WPE, CPEI, and AE) applied on the signal generated in (A).

and $C_i^m(r)$ is defined using the heavy side function $\Theta(u)$ (1 for $u > 0$, 0 otherwise) and a distance measure dist:

$$C_i^m(r) = \frac{\sum_{j=1}^{N-(m-1)\tau} \Theta\left(r - \mathrm{dist}\left(X_i^{m,\tau}, X_j^{m,\tau}\right)\right)}{N - (m-1)\tau} \tag{10}$$

Here, the value of r is set to be 0:2 times the data standard deviation as per the thorough discussion in Chon et al. (2009). The distance measure we use is the same suggested in Pincus (1991) and can be formulated as:

$$\mathrm{dist}\left(X_i^{m,\tau}, X_j^{m,\tau}\right) = \max_{k=1,\ldots,m} \left|x_{i+(k-1)\tau} - x_{j+(k-1)\tau}\right|$$

The composite PE index (CPEI) is an alteration of permutation entropy that differentiates between the types of patterns. It is calculated as the sum of two permutation entropies corresponding to motifs having different delays where the latter (denoted as τ in this chapter) is determined by whether the motif is monotonically decreasing or increasing. CPEI, which we denote by H_i in this chapter, responds rapidly to changes in EEG patterns and can be defined as follows (Olofsen et al., 2008):

$$H_i = \frac{1}{\ln(m! + 1)} \frac{H(m, 1) + H(m, 2)}{2} \tag{11}$$

The normalization denominator in Eq. (11) consists of the original number of motifs in addition to a newly introduced motif to account for ties (ties describe cases where negligible differences in amplitude occur within a motif). As a side note, the averaging step performed in that equation is highly approximate because of the lack of independency between motifs at different delays.

It is noticeable that WPE consistently drops for portions of the signal showing pulses. This is desired because of the lesser complexity of these regions and expected because of their immunity to noise. Here, we assume that the information contained in the examined signals is amplitude-dependent. Such results meet our expectations because WPE is clearly able to differentiate between burst and stagnant regions of the pulse train. In other words, using the variance contributes to weakening the noise effects and assigning more weight to the regular spiky patterns corresponding to a higher amount of information, which results in easier predictability and less complexity. It is important to note two things: (1) the contribution of patterns with higher variance toward the value of WPE dominates those of patterns with lesser variance, which highlights the powerfulness of the method in detecting abrupt changes in the input signal, and (2) the fact that WPE is computed within a specific time window explains why WPE values corresponding to impulsive segments of the signal do not decrease in spite of the decreasing amplitudes of the spikes (the normalization effect in Eq. 4 takes place within each window). We also plot in Fig. 4 the values of PE and WPE for different levels of signal-to-noise ratio (SNR). As anticipated, both entropy measures decrease with the increase of the SNR because the effect of noise contributing to more complexity becomes less significant. WPE decreases at a higher pace than PE, which reflects a better robustness to noise. As a final note on this section, we point out that traditional methods such as zero-crossing spike detection techniques might be useful for the purpose of this simulation; however, the sought goal was to demonstrate using synthetic data the ability of WPE to discriminate between regimes of data.

3.2 Single-Channel EEG Data Analysis

In Fig. 5, the same comparisons are performed for a sample EEG recording processed as in Fadlallah et al. (2011). High-pass filtering was further applied on the signal because we are

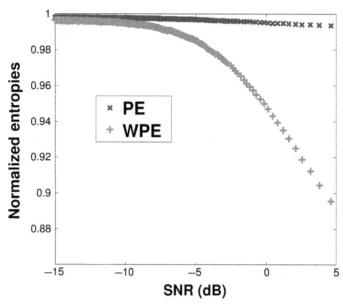

Fig. 4 Normalized PE and WPE values for different SNR levels. The signal used is the same as in Fig. 3.

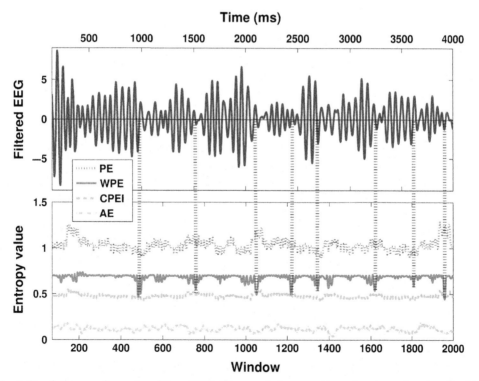

Fig. 5 Simulations performed on filtered EEG data sampled at 1000 Hz and processed as described in Fadlallah et al. (2011). WPE outperforms other entropy measures in location regiments exhibiting abrupt changes in the signal. The window length used for this plot was 114 with an overlap of two samples.

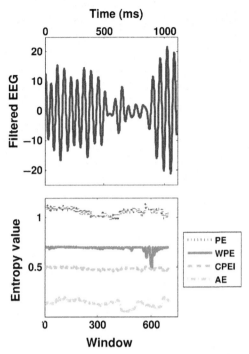

Fig. 6 Same procedure applied on a processed EEG portion corresponding to another channel. WPE mirrors best the sharp change in the signal noticeable before $t = 850$ ms. The window size used was 200 with an overlap of two samples at each iteration.

interested in removing very low-frequency components. It can be seen that WPE locates the regions where abrupt changes occur in the initial signal more accurately than the other methods, which is in line with our original expectations.

The same is reflected in Fig. 6 that shows a processed EEG portion corresponding to another channel. Our simulations show that increasing m beyond 4 affects the running time without significantly changing the obtained entropies. This is in line with the findings in Staniek and Lehnertz (2007)) where the parameter selection problem has been addressed in Olofsen et al. (2008). For situations where the effect of m is more pronounced, the running time issue can be addressed by speeding up the sliding of the window, as this entails a higher number of affected patterns at each instance.

Two examples of possible m-dimensional vectors corresponding to the same motif. The value of m used is 3.

3.3 Multichannel EEG Data Analysis

3.3.1 Setting

We propose to tackle the problem suggested in Fadlallah et al. (2011) from the perspective of the method presented above. The experimental setting exploits the steady-state

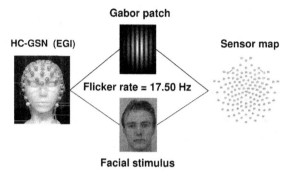

Fig. 7 Experimental setting using a HydroCell Geodesics Sensor Networks system (Electrical Geodesics, Inc., 2007).

visual evoked potential (ssVEP) paradigm by flashing a visual stimulus at a rate of 17:5 Hz to a participant. Two types of stimuli were presented to the subject, one representing an image of a neutral human face and the second a Gabor patch (Fig. 7). Each stimulus was presented for 4:2 s (plus 0:4 s prestimulus baseline). A surface Laplacian method was applied on the raw EEG data and the experiment's objective was to identify the active regions involved in the cognitive processing of each stimulus and study the corresponding connectivity patterns between all channel locations. In their work, Fadlallah et al. (2012) demonstrated that two traditional coupling methods (Pearson's correlation and mutual information) and one novel approach termed "generalized measure of association" (or GMA) were used to calculate bivariate interactions with respect to a single parieto-occipital channel chosen as reference (channel POz in a standard 10-20 referential configuration). Dependence values were computed per time windows of 114 samples.

The rationale for choosing this specific time window can be summarized as follows: the selected window size should (1) allow tracking the signal behavior with high time resolution, that is, using a reduced number of samples, (2) include enough samples that allow the estimation of permutation entropy quantities, and (3) relate to the observed physiological properties of the cognitive system being studied. Setting the window size to 114 verifies the three conditions (because 114 samples correspond to two periods of the ssVEP signal and roughly match the propagation time between brain cortices). Using this setting, higher coupling was observed for the face condition between occipital sites and the temporal-parietal-occipital sites neighboring P 4. The methodology suggested in this chapter will be applied on the same experimental data to infer functional relationships across different electrode sites.

3.3.2 Procedure

A precursor for a useful usage of PE (WPE) within the above context is to assign a "complexity" curve for each recorded signal, corresponding to an array of PE (WPE) values computed over a given time window (114 ms in this case). We can then compute the dependence between the different channels by simply applying correlation on these curves. Intuitively, this implies using a linear measure of dependence to measure how close the complexity of two time series is. In our simulations, we select Spearman's rho as a measure of statistical dependence between the different PE (WPE) curves. In Fig. 8, the obtained correlation values are mapped onto the corresponding locations on the human scalp. In the case of WPE (Fig. 8A and B), more activity can be spotted in locations that seem to point toward sources in the occipito-parieto-temporal area of the right brain hemisphere. This outcome aligns with the results obtained in Fadlallah et al. (2012), which, as previously mentioned, indicate higher activity in that specific region. On the other hand, PE tends to show activity localized in right posterior areas.

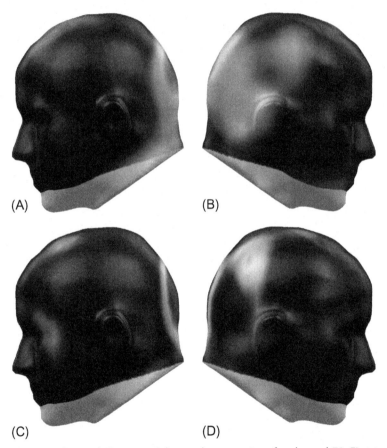

(A)   (B)

(C)   (D)

Fig. 8 Using Spearman's correlation to weight graph connections for channel 72. First two subplots (A and B) show interpolated correlation measures over right and left (R and L) head surface for the face condition (F) when using WPE and subsequent subplots (C and D) exhibit the same when using PE. A statistical assessment of the discriminatory performance between the two conditions can be seen in Fig. 9 and Table 1.

3.3.3 Statistical Analysis

We use the two-sample Kolmogorov-Smirnov (KS) test applied on the obtained distributions with the null hypothesis being that the two samples are drawn from the same distribution. The KS test tries to estimate the distance between the empirical distribution functions of the two samples. Assuming $\gamma_1(x)$ and $\gamma_2(x)$ to be the sample vectors, it can be calculated as $K_{\gamma_1, \gamma_2} = \max_x |F_{\gamma_2}(x) - F_{\gamma_1}(x)|$ where $F_{\gamma 1}(x)$ and $F_{\gamma 2}(x)$ denote the empirical cumulative distribution functions for the n iid observations, alternatively $F_{\{X_1, \ldots, X_n\}}(x) = \frac{1}{n}\sum_{i=1}^{n} I_{X_i \leq x}$ where I_k denotes the indicator function. The null hypothesis is rejected at the α-level if $\sqrt{(n_1 n_2)/(n_1 + n_2)}\, S_{\gamma_1, \gamma_2} > K_\alpha$ where n_1 and n_2 denote the number of samples from each observation vector and K refers to the Kolmogorov distribution (Chon et al., 2009). In our case, $n_1 = n_2 = 45$ and $\alpha = 0.05$.

3.3.4 Discussion

Fig. 9 and Table 1 show that, unlike PE, WPE is able to discriminate the two conditions with a statistically significant KS test. A possible explanation is the inconsistency in PE's tracking of steep changes in the processed signals, which creates ad hoc dependencies when computing the pairwise correlations and results in the indiscernibility of the two conditions. This problem is avoided when using WPE because the latter follows faithfully the change trends in the signal, as illustrated in Fig. 3.

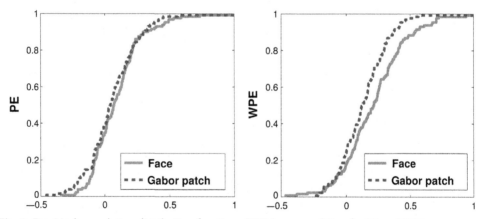

Fig. 9 Empirical cumulative distribution functions (CDFs) per condition for PE and WPE.

Table 1 Two-sample Kolmogorov-Smirnov test results

KS test	PE	WPE
Null hypothesis rejection	False	True
p-Value	0.508	0.009
Test statistic	0.101	0.202

3.4 Epilepsy Detection

3.4.1 Setting

Next, we propose to apply WPE for epilepsy detection. We use the same data as Quiroga et al. (1997, 2002), in which tonic-clonic seizures of a subject were recorded with a scalp right central electrode (located near C4 in a standard 10-20 montage). The recording consisted of 3 min, including around 1 min of preseizure time and 20 s of postseizure activity. A sampling rate of 102.4 Hz was used to collect the signal.

3.4.2 Discussion

We computed different measures of entropy on windows of 50 samples of data slid by five samples (Fig. 10B). The obtained curves are further smoothed in Fig. 10C using a moving average filter of length 35 samples. The commencement of epileptic activity in the recorded signal induces noticeable changes for all entropy measures, in particular for

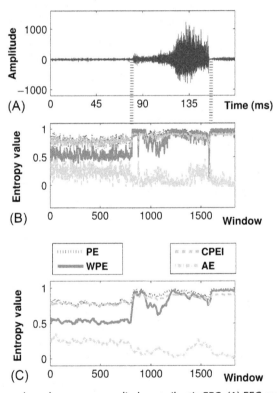

Fig. 10 Different entropy-based measures applied on epileptic EEG. (A) EEG recording of an epileptic subject. The recording, sampled at 102.4 Hz, contains approximately 1 min of preseizure activity and 20 s of postseizure activity. (B) Different measures of entropy computed using a sliding window of 50 sample with five samples overlap. (C) Smoothed entropy measures curves obtained by applying a moving average filter of length 35 samples.

WPE that exhibits a significant jump in value. This is further quantified by computing the ratio of average measured entropies of epileptic and nonepileptic segments (Table 2), which shows a more pronounced difference between both portions for WPE. The latter achieves almost twice better discriminability between the two portions of the signal, that is, 42% better than the next closest measure (CPEI).

4. CONCLUSION

This chapter proposes a different definition of permutation entropy that retains the amplitude information of nonlinear time series. A method to weight the motif counts by statistics derived from the signal patterns has been proposed. The new method is different from PE, however, in the sense that it suits better signals having considerable amplitude information. For the range of signals that do not verify this property, PE might be a better choice. Simulations were carried on spiky synthetic data and human EEG recordings that underwent narrow-band filtering, taking into account the variance of the mentioned patterns. The measure showed consistency when applied on various regions of both signals by differentiating distinct regimes and assigning similar complexities for analogous portions. Moreover, WPE decreases for higher SNRs, which corroborates the fact that noise has higher complexity. The suggested method was also applied on processed EEG data to differentiate two cognitive states as suggested in Fadlallah et al. (2012), with the help of the Kolmogorov-Smirnov statistical tool and epileptic data to detect seizure onset. The power of permutation entropy as a simple and computationally fast measure for time series complexity has been hence confirmed on both synthetic and real data. Future work includes analyzing more thoroughly the effect of the free parameters (m, noise model, etc.), exploring other weighting schemes, and comparing to other nonlinear regularity estimators based on equiquantal or equiprobable binning.

REFERENCES

Bandt, C., Pompe, B., 2002. Permutation entropy: a natural complexity measure for time series. Phys. Rev. Lett. 88. 174102.
Bruzzo, A.A., Gesierich, B., Santi, M., Tassinari, C.A., Birbaumer, N., Rubboli, G., 2008. Permutation entropy to detect vigilance changes and preictal states from scalp EEG in epileptic patients. A preliminary study. Neurol. Sci. 29, 3–9.
Cao, Y., Tung, W.W., Gao, J.B., Protopopescu, V.A., Hively, L.M., 2004. Detecting dynamical changes in time series using the permutation entropy. Phys. Rev. E. 70. 046217.
Chon, K.H., Scully, C.G., Lu, S., 2009. Approximate entropy for all signals is the recommended threshold value r appropriate? IEEE Eng. Med. Biol. 28, 18–23.
Commission on Epidemiology and Prognosis, International League Against Epilepsy, 1993. Guidelines for epidemiologic studies on epilepsy. Epilepsia 34, 592–593.
Daoming, Z., Guojun, T., Jifei, H., 2008. Fractal random walk and classification of ECG signal. Int. J. Hybrid Inf. Technol. 1, 1–9.
Electrical Geodesics, Inc., 2007. Geodesics Sensor Networks Technical Manual. Riverfront Research Park, Eugene, OR.

Fadlallah, B.H., Seth, S., Keil, A., Principe, J.C., 2011. Robust EEG preprocessing for dependence-based condition discrimination. IEEE Eng. Med. Biol., 1407–1410.

Fadlallah, B., Seth, S., Keil, A., Principe, J., 2012. Quantifying cognitive state from EEG using dependence measures. IEEE Trans. Biomed. Eng. 59, 2773–2781.

Fadlallah, B., Chen, B., Keil, A., Principe, J., 2013. Weighted-permutation entropy: a complexity measure for time series incorporating amplitude information. Phys. Rev. E. 87022911.

Graff, B., Graff, G., Kaczkowska, A., 2012. Entropy measures of heart rate variability for short ECG datasets in patients with congestive heart failure. Acta Phys. Pol. B: Proc. Suppl. 5, 153.

Guiasu, S., 1971. Weighted entropy. Rep. Math. Phys. 2, 165–179.

Hagmann, C.F., Robertson, N.J., Azzopardi, D., 2006. Artifacts on electroencephalograms may influence the amplitude-integrated EEG classification: a qualitative analysis in neonatal encephalopathy. Pediatrics 118, 2552–2554.

Hellström-Westas, L., Rosén, I., de Vries, L.S., Greisen, G., 2006. Amplitude-integrated EEG classification and interpretation in preterm and term infants. NeoReviews 7, e76–e87.

Kurths, J., Schwarz, U., Witt, A., Krampe, R.T., Abel, M., 1996. Measures of complexity in signal analysis. AIP Conf. Proc. 33–54.

Li, X., Ouyang, G., Richards, D.A., 2007. Predictability analysis of absence seizures with permutation entropy. J. Epilepsy Res. 77, 70–74.

Li, X., Cui, S., Voss, L.J., 2008. Using permutation entropy to measure the electroencephalographic effects of sevoflurane. Anesthesiology 109, 448–456.

Li, Z.H., Ouyang, G.X., Li, D., Li, X.L., 2011. Characterization of the causality between spike trains with permutation conditional mutual information. Phys. Rev. E. 84.

Litt, B., Echauz, J., 2002. Prediction of epileptic seizures. Lancet Neurol. 1, 22–30.

Lytton, W., 2008. Computer modelling of epilepsy. Nat. Rev. Neurosci. 9, 626–637.

McSharry, P.E., Smith, L.A., Tarassenko, L., 2003. Prediction of epileptic seizures: are nonlinear methods relevant? Nat. Med. 9, 241–242.

Minasyan, G.R., Chatten, J.B., Chatten, M.J., Harner, R.N., 2010. Patient-specific early seizure detection from scalp electroencephalogram. J. Clin. Neurophysiol. 27, 163–178.

Olofsen, E., Sleigh, J.W., Dahan, A., 2008. Permutation entropy of the electroencephalogram: a measure of anaesthetic drug effect. Brit. J. Anaesth. 101, 810–821.

Pincus, S.M., 1991. Approximate entropy as a measure of system complexity. Proc. Natl. Acad. Sci. U. S. A. 88, 2297–2301.

Quiroga, R.Q., Blanco, S., Rosso, O.A., Garcia, H., Rabinowicz, A., 1997. Searching for hidden information with Gabor Transform in generalized tonic-clonic seizures. Electroencephalogr. Clin. Neurophysiol. 103, 434–439.

Quiroga, R.Q., Garcia, H., Rabinowicz, A., 2002. Frequency evolution during tonic-clonic seizures. Electromyogr. Clin. Neurophysiol. 42, 323–331.

Shah, D.K., Mackay, M.T., Lavery, S., Watson, S., Harvey, A.S., Zempel, J., Mathur, A., Inder, T.E., 2008. Accuracy of bedside electroencephalographic monitoring in comparison with simultaneous continuous conventional electroencephalography for seizure detection in term infants. Pediatrics 121, 1146–1154.

Stafstrom, C., 2006. Epilepsy: a review of selected clinical syndromes and advances in basic science. J. Cereb. Blood Flow Metab. 26, 983–1004.

Staniek, M., Lehnertz, K., 2007. Parameter selection for permutation entropy measurements. Int. J. Bifurcat. Chaos 17, 3729–3733.

Tao, J.D., Mathur, A.M., 2010. Using amplitude-integrated EEG in neonatal intensive care. J. Perinatol. 30, S73–S81.

World Health Organization, 2012. Fact Sheet No. 999. .

Zunino, L., Zanin, M., Tabak, B.M., Perez, D.G., Rosso, O.A., 2009. Forbidden patterns, permutation entropy and stock market inefficiency. Physica A 388, 2854–2864.

FURTHER READING

George, M., Wan, T.W., Jingbo, W., 2003. Evaluating Kolmogorov's distribution. J. Stat. Softw. 8, 1–4.

CHAPTER 7

Biological Knowledge Graph Construction, Search, and Navigation

Chandana Tennakoon*, Nazar Zaki*, Hiba Arnaout[†], Shady Elbassuoni[†], Wassim El-Hajj[†], Alanoud Al Jaberi*

*College of Information Technology, United Arab Emirates University, Al Ain, United Arab Emirates
[†]Department of Computer Science, American University of Beirut, Beirut, Lebanon

1. RESOURCE DESCRIPTION FORMAT (RDF) KNOWLEDGE GRAPHS

In recent years, the web has evolved from a network of linked documents to one where both documents and data are linked, resulting in what is commonly known as the Web of Data. Underpinning this evolution is a set of best practices known as Linked Data,[1] which provides mechanisms for publishing and connecting structured data on the web in a machine-readable form with explicit semantics. The increasing adoption of Linked Data is turning the web into a global dataspace that connects data from diverse domains and enables genuinely novel applications. Linked Data has grown from an academic endeavor into one that has been embraced by numerous governments and industrial stakeholders in domains such as business and finance, geography, governance, media, digital libraries, and life sciences.

Linked data is typically represented as sets of resource description framework (RDF)[2] triples. An RDF triple is a triple with three fields: a *subject*, a *predicate*, and an *object*, where subjects and predicates are uniform resource identifiers (URIs) and objects are either URIs or literals. A few sample triples from the Wikipedia-based RDF dataset DBpedia (Auer et al., 2007) are shown in Table 1. DBpedia is a general-purpose triple store[3] that has transformed Wikipedia entries to triples.

Alternatively, an RDF dataset can be also viewed as a labeled graph, which is commonly referred to as a knowledge graph (KG). In a knowledge graph, node labels are *either* URIs representing resources *or* literals, and edge labels are URIs representing predicates. Fig. 1 shows the subgraph corresponding to the triples in Table 1. We omit the prefix of all URIs for readability.

---

[1] http://linkeddata.org.

[2] http://www.w3.org/RDF.

[3] A triple store or RDF store is a purpose-built database for the storage and retrieval of triples through semantic queries.

Leveraging Biomedical and Healthcare Data
https://doi.org/10.1016/B978-0-12-809556-0.00007-1

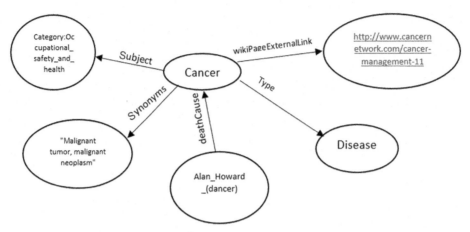

Fig. 1 RDF knowledge graph representation. A subgraph depiction corresponding to the RDF triples shown in Table 1, where every node represents either a resource (URI) or a literal (string), and every edge represents another resource (URI) that is called a relationship.

Table 1 RDF Triples

Subject	Predicate	Object
http://dbpedia.org/ resource/Cancer	http://purl.org/dc/ terms/subject	http://dbpedia.org/resource/ Category:Occupational_safety_ and_health
http://dbpedia.org/ resource/Cancer	http://dbpedia.org/ property/synonyms	"malignant tumor, malignant neoplasm"
http://dbpedia.org/ resource/Cancer	http://dbpedia.org/ ontology/ wikiPageExternalLink	http://www.cancernetwork.com/ cancer-management-11
http://dbpedia.org/ resource/Cancer	http://www.w3.org/ 1999/02/22- rdf-syntax-ns#type	http://dbpedia.org/ontology/ Disease
http://dbpedia.org/ resource/Alan_ Howard_(dancer)	http://dbpedia.org/ ontology/deathCause	http://dbpedia.org/resource/ Cancer

The semantic query language for RDF is known as SPARQL. It allows composing structured queries consisting of triple patterns and conjunctions. In general, a SPARQL query Q consists of n-triple patterns, where a triple pattern is a *subject-predicate-object (SPO) triple* with variables. A variable occurring in one of the triple patterns can be seen again in another triple in the same query Q, denoting a join condition. Consider a query to find the names of actors who died from cancer. The query would be written as follows.

< ?x; deathCause; Cancer
?m; starring; ?x >

In this query, *x* denotes the name of the actor and *m* denotes the name of a movie he/she starred in. Two possible answers for the above query are:

< Paul_Newman; deathCause; Cancer
The_Verdict; starring; Paul_Newman >

< Alan_Rickman; deathCause; Cancer
Love_Actually; starring; Alan_Rickman >

Although SPARQL gives very specific results, the user should know how to write the query entities and put the variables in a proper way to get any results at all. For this, he/she must be aware of the vocabulary and format of the underlying data.

1.1 Biomedical Data on the Web

Today, there is a wealth of biological data on the web. For instance, there are almost 1900 publicly available biological databases (Galperin et al., 2017) as well as many private ones. However, such data is typically very diverse in scope and format. For instance, some of these datasets are available as delimited text files, VCF, BED, GFF, or Excel format. Similarly, many popular databases such as those at NCBI[4] are available only in tab or comma-separated format. This lack of standard data format hinders the effectiveness of information retrieval from such data sources. With the rise of RDF as a standard format for sharing and publishing data on the web, there is a current trend in the biomedical domain to publish biomedical data in RDF format as well (Zaki et al., 2017). For example, UNIPROT (The UniProt, 2015) is one such effort. UNIPORT is a publicly available database of protein sequence and function information that contains around 13 billion RDF triples.

In addition to providing a common format for publishing the data, RDF enables these different datasets to be seamlessly interconnected and linked. This will allow users to retrieve information that might be spread across multiple knowledge bases, which will increase the accessibility of such data sources and truly enable addressing complex knowledge quests that cannot be addressed using just one data source.

1.1.1 Biomedical RDF Knowledge Graphs

There have been many efforts aimed at making heterogeneous biomedical data available in RDF format. In Ermilov et al. (2013), the authors attempted to convert tabular data such as CSV files or Excel sheets to RDF. They applied their approach to 10,000 datasets and ended up with 7.3 billion triples. The data was generic in nature and not restricted to the biological domain. A similar but older work (Reck, 2003; Grove, n.d.) was also carried out by the Maryland Mindswap Lab. Another work (Han et al., 2008) developed two conversion tools. The first is a graphical interface that allows the user to create and

---

[4] www.ncbi.nlm.nih.gov.

remove a vertex/edge, drag a vertex, and change properties of a vertex/edge. The second is a web application that takes as input a URL to a Google spreadsheet or a CSV file and an RDF template, and provides RDF triples as output. Another recent effort in this area is the work in Chiba et al. (2015), which only focused on genomic data and Bio-Carian[5] (Zaki and Tennakoon, 2017), where tabular biological data was converted to RDF and a faceted search engine was developed to search the data. Bio2RDF (Belleau et al., 2008) is another proposed approach to connect diverse biological databases. It converts data from different sources to RDF. Moreover, it connects the data together in a knowledge base following a set of defined rules. Linked life data or LLD (Momtchev et al., 2009) is another approach that transformed and connected more than 20 biological datasets. This is accomplished by defining a set of rules and patterns (e.g., naming conventions, inferring relationships).

Another popular knowledge graph for biomedical sciences is KnowLife (Ernst et al., 2015). The construction of this knowledge graph started with seed facts and relations, a dictionary of biomedical entities, and a text corpus. The input data must be processed by following a few stages. The data is first submitted for entity recognition, pattern gathering and analysis, and finally, for consistency reasoning before being represented in the form of an RDF graph. Another work that enables biological data integration is KaBoB (Livingston et al., 2015). The authors of KaBoB provide a knowledge base of biomedicine that is constructed from 18 biomedical databases using representations grounded in Open Biomedical Ontologies. A similar effort is the ontology-based integrated database hub, the BioLOD (Toyoda, 2012). The authors were able to make almost 7000 OWL/RDF files available for the public.

Another major challenge that needs to be addressed to increase the usability of biomedical data is the development of a scalable explorative search engine to explore such data. Many research efforts addressed this problem by proposing browsers, with different methods of browsing, for searching RDF data. One method is using a faceted search (Zaki and Tennakoon, 2017; Huynh and Karger, 2009; Kobilarov and Dickinson, 2008), where the search can start with SPARQL or free-text query, and then a set of facets and facet values can be given to the user for further exploration. Another method for browsing is graph navigation (Philipp Heim and Steffen, 2008; Russell, 2008; Schweiger et al., 2014; e Zainab et al., 2015), for the purpose of visualizing the data.

2. OUR PROPOSED APPROACH TO CREATE, SEARCH, AND VISUALIZE BIOLOGICAL KNOWLEDGE GRAPHS

In this section, we will present one recent approach that addresses the different challenges discussed above, starting with the construction of a rich and reliable biological knowledge

---

[5] http://www.biocarian.com/.

graph, providing a live search service using free-text queries, and finally displaying the graph and allowing the user to explore it in more depth.

2.1 Creating the Knowledge Graph

Starting with a set of biological databases, we can identify several basic entities, where each basic entity belongs to a biological layer that we define:

1. Genomic layer: Data related to the genomic structure
   Example: sequencing data, structural variation data.
2. Transcription processes: Data related to gene expression
   Example: genes, RNA, transcripts.
3. Proteins: Data related to proteins
   Example: proteins, protein structure.
4. Biological processes: Data related to biological processes
   Example: biological pathways, drug information, diseases.

For each layer, out of the four layers defined above, a number of entities can be defined. For example, single-nucleotide polymorphism (SNPs), proteins, and biological pathways are three such entities. Each basic entity is usually associated with one or more databases that summarize the facts known about it. Several entity examples, with their key databases and defined layers, are shown in Table 2. A basic entity will be called a "canonical entity" and its associated database will be called a "canonical database."

Next, we need to prepare a knowledge graph that connects these heterogeneous databases. The obvious step is to start with the most trusted information, namely the canonical entities. The canonical entities are given types. This annotation can be represented using the standard "rdf:type" predicate. Moreover, we will have a reserved namespace to avoid clashes with other type information and unique types to describe each canonical entity. In Fig. 2, we show two examples of canonical entities, the first is a straightforward annotation, where the entity is a URI; the second is a literal dealt with using an indirect annotation. In a biological knowledge graph, both a subject and an object can be a canonical entity. The subject in the following triple must be annotated as a canonical entity:

< dbSNP:10177chr1; rdf:description; "rs367896724">

dbSNP:10177chr1 is an SNP object that describes the SNP at chr1:10177. A type edge is created in this case to annotate the entity:

<dbSNP:10177chr1; rdf:type; canonical:SNP>

This simple addition cannot be made, however, in the following case. One of the knowledge graph rules states that a literal cannot be a subject. For example, in the triple:

< GWAS:breast_cancer; GWAS:SNP; "rs10069690">

the object "rs10069690" is the name of the SNP associated with breast cancer. Because it is not possible to attach an edge to a literal in RDF, we will have an alternative for the literal object, namely a blank node, and then attach the canonical annotation with the blank node created:

Table 2 Canonical entities and canonical databases

Canonical entity	Canonical database	Biological layer
SNP	dbSNP (Sherry et al., 2001)	Genomic
Protein	UNIPROT (The UniProt, 2015)	Proteins
Biological pathways	KEGG (Kanehisa et al., 2016)	Biological processes
Disease	OMIM (Amberger et al., 2015)	Biological processes
Biological pathways	KEGG (Kanehisa et al., 2016)	Biological processes

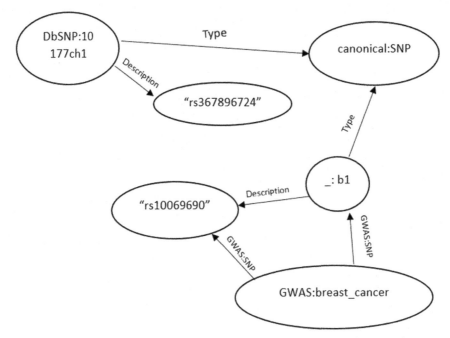

Fig. 2 Annotating canonical entities in the knowledge graph. Two examples of canonical entities: a URI and a literal, where a solution is offered for the case of a literal subject, which is usually not allowed in any standard knowledge graph.

   <GWAS:breast_cancer; GWAS:SNP; _:b1>
   <_:b1; rdf:description; "rs110069690">
   <_:b1; rdf:type; canonical:SNP>

After annotating the canonical entities, we need to associate them with their canonical databases. For this, we can make use of the standard IDs that are usually associated with the canonical entities. For example, an entity of type SNP has a dbSNP identifier, an entity of type protein has a UNIPROT identifier, and an entity of type gene has HGNC and Ensemble identifiers. Notice that the last case has two identifiers. We can process this by transforming one to another. The canonical databases will have their unique namespaces. Therefore, a standard ID for an entity will be stored under the URI DATA-BASANAME:ID. This unique URI is called a canonical URI. In other words, every canonical entity's standard ID will be associated with its corresponding canonical URI, using the predicate "rdfs:sameAs." This fulfills the purpose of creating a path between the canonical entity, its standard ID, and its canonical URI. This path will allow a user to access a database's canonical entity to explore more knowledge about it. To illustrate this concept, let's consider the following example, a canonical entity A of type protein has a standard ID MAPK3. This standard ID can also be described using a UNI-PROT ID P27361, that is, a canonical database ID. This infers the canonical URI UNI-PROT:P27361, which will be associated with the canonical entity A using the predicate "rdfs:sameAs." More specifically, starting with the following two triples:

< A; rdf:description; "MAPK3">

and

< A; rdf:type; canonical:PROTEIN>

we can infer the following triple:

< A; rdfs:sameAs; UNIPROT:P2731>

An illustration of this example is shown in Fig. 3.

Canonical databases will form a strong base for our knowledge graph and trusted sources of information. However, for the knowledge graph to be richer, we need to extend it within and outside the database. This is accomplished by adding new edges labeled "rdfs:seeAlso." More specifically, any two subjects that are connected with the same canonical database entry will be linked using a "rdfs:seeAlso" predicate. This way, the user can go from the first subject S1 to the second subject S2, and vice versa. To further extend the knowledge graph, we can infer new possible connections. This can be expressed using a "rdfs:inferred" predicate. If S1 AND S2 are two subjects that are connected by a set of paths P1,..,Pn, S1 and S2 will be connected using:

< S1; rdfs:seeAlso; S2>

and

< S2; rdfs:seeAlso; S1>

The user can now follow the inferred link to find more information that might be related to his/her quest. The challenge in this inference is that it must be made based on some sort of logical decision, and then even after that, it can be a valid connection or not. For example, Fig. 4 shows four canonical entities: an SNP, a gene, a disease, and a pathway. It is known that the SNP affects the gene and it is inside a gene, and it is also known to be the cause of a disease. Moreover, the pathway also affects the disease. Based on the existing connections, one can find a path between the gene and the pathway, hence creating a connection between them. However, this connection might not be true. The gene is not necessarily connected to the pathway. It is possible that the SNP affects the disease through a different pathway. For this, a reliability or a confidence score can be very helpful to decide whether a suggested connection is valid. One measure can be the length of

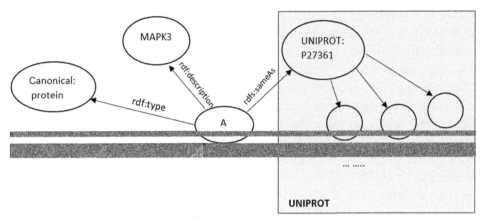

Fig. 3 Connecting canonical entities to canonical databases. An illustration of building a collection between a canonical database, which can be very accurate and trusted but very limited, and outside information by looking for canonical entity outside (i.e., in other databases).

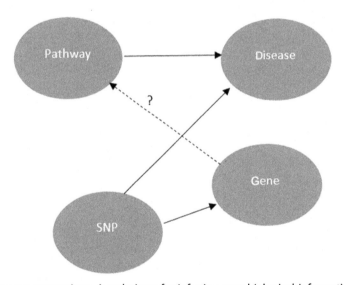

Fig. 4 Inferring new connections. A technique for inferring new biological information, hence new knowledge graph triples, by following a number of biological rules and criteria.

the path between the two nodes in question (i.e., the shorter the path, the better). If the connection has been made on this, another measure can help in deciding the reliability of an inferred connection. A score can be calculated using an algorithm such as PEWCC (Zaki et al., 2013). It will determine the reliability of an edge with a score using weights assigned to neighboring edges. The user can check the scores and decide by himself/herself to believe or follow the new inferred information or not.

2.2 Searching the Knowledge Graph

Constructing the knowledge graph based on the steps discussed above will result in a very large number of extracted and inferred triples. If the user wants to set up the knowledge graph in a standard desktop system, he/she will not find it practical. The best way to boost the efficiency of the searching process is to use live searching. A smaller portion of the data needs to be set up; however, the rest of the data can be loaded as needed from one or more external sources. Most of the canonical databases have APIs for accessing the data. They can be called after processing the user's query, and the data retrieved will then be converted based on the rules discussed for the knowledge base construction.

The easiest way for a user to search RDF data is by using a free-text query. Free-text search is done by creating an inverted index of the content in the database, or in this case, in the knowledge graph and the data associated with it. Most of the data in the knowledge graph is in the form of URIs. The goal is to create out of it texts that can be comparable with the free-text query of the user. The indexing process is presented in Fig. 5.

It is crucial to note that not all data require indexing. Some indexed data will occupy a sizable storage location but might not be needed at all. Therefore, the first step is to decide on the information that must be indexed. For example, it might not be important to index numerical quantities. This can be defined easily using the predicates that involve numerical data as the object of the triple. We then extract all the other triples (S,P,O) to be indexed. Every triple P will be associated with a text description, something that a user might enter. This description is called H(P), where H stands for *human-readable*.

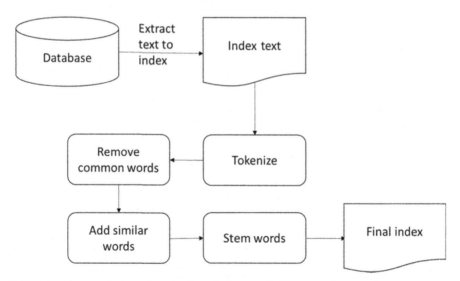

Fig. 5 Creating the text index using the knowledge graph. The steps for transforming a structured knowledge graph into an indexed text, so it becomes much easier for the system to match the user's text query to the content of the knowledge graph.

Associating the descriptions can be done manually, using the crowd (biology students for example), or extracted from the database itself (column headers for example). Next, every three-part fact (S,P,O) will be transformed into a two-part fact that contains a structured part and an unstructured part (S, H(P)+O). The first part is a URI representing the subject, and the second part is a free text part containing the human-readable description and the object of the original triple. Please note that the object of an RDF triple can either be a URI or a literal.

The key of the new representation of the triple is going to be the subject. The subject can exist in different triples with different objects. For every subject, we accumulate all the associated texts and objects. For example, the facts (S,T1), …, (S,Tn) are going to be replaced with (S,T1+⋯+Tn), where P is the key subject (a URI) and T1+⋯+Tn is the concatenation of all the objects and texts associated with the subject S. More specifically, consider the following triples:

<protein:BRCA1; proteing:dis; "breast cancer">

and

<protein:BRCA1; protein:chr; "17">

The first step is to replace the predicates with their descriptions, so:

<protein:BRCA1; "related to disease"; "breast cancer">

and

<protein:BRCA1; "is in chromosome"; "17">

The next step is to concatenate the predicate description with the object of the triple, leading to a two-part fact:

<protein:BRCA1; "related to disease breast cancer">

and

<protein:BRCA1; "is in chromosome 17">

The final step is to merge all the facts having the same key subject:

<protein:BRCA1; "related to disease breast cancer is in chromosome 17">

At this point, the necessary data (the data that will be indexed) is ready in the form of (S, T). The inverted index can be done using the popular open-source Solr (Smiley and Pugh, 2009) platform. For every fact, only the T part will be indexed. S is going to be the ID of the entry, and T will be the TEXT of the entry. During indexing, every entry will follow several steps. The first step is to tokenize the TEXT, so that the TEXT is transformed into a set of keywords where the delimiter is the white spaces. Next, the keywords will be reduced into a set of unique terms. Then, every term will be associated with its synonyms. This step can be helpful in adding support to translating the standard and canonical IDs used in the databases. For example, a possible synonym for the entrez gene ID is its corresponding HGNC ID. Another example is associating terms with alternative terms, "carcinoma" and "cancer." For this, we can use the crowd or automate the process using extensive thesauruses such as MESH (Dhammi and Kumar, 2014). The final step is to stem the terms. This is where the terms are converted to a form that allows free-text searching of different verb and noun forms. For example, if the user enters the words "menstruation" or "menstruating," he/she will be directed to the stemmed word "menstruat" in the indexed data.

Next, we need to define a way of processing the user's query and comparing it with the data we have. The user enters a free-text query that will be formatted according to the syntax compatible with Solr. Solr will search the TEXT fields of all the facts to find a match and return the corresponding key (ID) as a result for the search. The IDs retrieved will lead the system to their corresponding nodes in the knowledge graph built. Finally, the nodes will be displayed.

2.3 Displaying the Knowledge Graph

The final step in matching the user's query with the inverted index is to get the set of result nodes (IDs). The next challenge is to figure out how to display them, given that the nodes will probably be isolated in the knowledge graph. Another challenge is to allow the user to make further explorations in the knowledge graph. We will not take the approach of showing all possible connections to a node as it might display too much information, which might overwhelm the user. We will provide the user with nodes connected by a path of length L to a hit, where L is a user-defined value. If a connected path has the label rdfs:sameAs, rdfs:seeAlso, and rdfs:inferred, we will not follow the paths further. The user will not have a full view of the connections between nodes added during the processing of the database. The user will have the connections with the highest confidence score in the knowledge graph.

The result knowledge subgraph might contain some artificially added connections, based on the ideas discussed in the construction phase:

- "rdfs:sameAs": shows additional information about the node entity.
- "rdfs:seeAlso": shows other nodes related to the node entity.
- "rdfs:inferred": shows nodes that might have a relationship with the node.

Now, choosing to follow these newly created connections is left for user discretion, based on weighted scores (reliability, confidence, etc.) given by the search engine. When the user selects to follow a certain path by clicking on a specific node, the same procedures are applied, focusing on the selected node. After a few graph navigation rounds, the user might lose track of the search. He/she might want to go back and forth in the graph. To solve this problem, the search history will be stored so that the user can move freely in his/her search.

Consider a query where the user is asking to find SNPs in a gene G that are related to diseases. The query can be answered as follows:

1. Find a set of entities of type disease (set of entities D) that are related to the entity of type gene (entity G).
2. Find a set of entities of type SNP (set of entities S) that are related to diseases in D.

3. Find SNP entities from S that are related to the gene entity G.

The intuition behind the navigation logic is to give the user just enough data about his information need. More specifically, the goal is to give the user all the entities that are related to the query without overwhelming him/her with too much information. An overall view of the search engine is given in Fig. 6. The model starts with several biological databases, and it transforms them (if needed) to RDF. Then, RDF can be represented as a graph. Next, the graphs representing the databases are interconnected based on some rules (standard IDs and canonical IDs). The graphs are also indexed for the purpose of allowing free-text searches. Finally, the text search is accessed through a browser interface.

3. EXTENDING THE KNOWLEDGE GRAPH USING MACHINE-LEARNING TECHNIQUES

We can also explore the possibility of incorporating machine-learning methods to predict additional confident links. We can observe in many situations that similar entities have a higher chance of being related. This property is called *homophily* and it is possible to derive several measures of similarity between entities using the observable features in the graphs, that is, the neighborhood of nodes and the paths between nodes. We can use either the local features of the graphs or global features. The latter tends to give a better idea

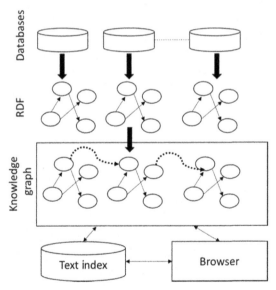

Fig. 6 Creating, searching, and navigating a biological knowledge graph. A brief overview of the system that creates a biological RDF knowledge graph out of multiple heterogeneous biological databases, provides a user friendly method for searching using free text, and allows the user to navigate and explore the result subgraph by asking for more information.

of similarity but will be more time consuming. One solution is to have compromise quasilocal methods that use a subset of global information and also rely on graph-based information. These methods will generate a set of features for each pair of the related nodes. With the features calculated for node pairs, we can use standard machine-learning algorithms to extract further relationships.

4. CONCLUSION

Scientists often need to access several biological databases to get a full answer for their research needs. In this chapter, we have shown how these databases can be effectively and efficiently converted into an RDF knowledge graph. We discuss a possibility of inferring new information from existing data, and how to score them based on a reliability measure. Moreover, we showed how the knowledge graph can be accessed using free-text search.

REFERENCES

Amberger, J.S., Bocchini, C.A., Schiettecatte, F., Scott, A.F., Hamosh, A., 2015. OMIM.org: Online Mendelian Inheritance in Man (OMIM($^{®}$)), an online catalog of human genes and genetic disorders. Nucleic Acids Res. 43, D789–D798.

Auer, S., Bizer, C., Kobilarov, G., Lehmann, J., Cyganiak, R., Ives, Z., 2007. In: Aberer, K., Choi, K.-S., Noy, N., Allemang, D., Lee, K.-I., Nixon, L., Golbeck, J., Mika, P., Maynard, D., Mizoguchi, R., Schreiber, G., Cudré-Mauroux, P. (Eds.), DBpedia: a nucleus for a web of open data. Proceedings of the Semantic Web: 6th International Semantic Web Conference, 2nd Asian Semantic Web Conference, ISWC 2007 + ASWC 2007, Busan, Korea, November 11–15. Springer Berlin Heidelberg, Berlin, Heidelberg, pp. 722–735.

Belleau, F., Nolin, M.-A., Tourigny, N., Rigault, P., Morissette, J., 2008. Bio2RDF: towards a mashup to build bioinformatics knowledge systems. J. Biomed. Inform. 41, 706–716.

Chiba, H., Nishide, H., Uchiyama, I., 2015. Construction of an ortholog database using the semantic web technology for integrative analysis of genomic data. PLoS One. 10, e0122802.

Dhammi, I.K., Kumar, S., 2014. Medical subject headings (MeSH) terms. Indian J. Orthop. 48, 443–444.

e Zainab, S.S., Hasnain, A., Saleem, M., Mehmood, Q., Zehra, D., Decker, S., 2015. In: FedViz: a visual interface for SPARQL queries formulation and execution. VOILA at ISWC.

Ermilov, I., Sören, A., Claus, S., 2013. Csv2rdf: User-driven csv to rdf mass conversion framework. In: Proceedings of the ISEM, vol. 13, pp. 4–6.

Ernst, P., Siu, A., Weikum, G., 2015. KnowLife: a versatile approach for constructing a large knowledge graph for biomedical sciences. BMC Bioinformatics 16, 157.

Galperin, M.Y., Fernández-Suárez, X.M., Rigden, D.J., 2017. The 24th annual nucleic acids research database issue: a look back and upcoming changes. Nucleic Acids Res. 45, D1–D11.

Grove, M., n.d. Mindswap Convert To RDF Tool. http://www.mindswap.org/~mhgrove/convert/.

Han, L., Finin, T., Parr, C., Sachs, J., Joshi, A., 2008. In: Sheth, A., Staab, S., Dean, M., Paolucci, M., Maynard, D., Finin, T., Thirunarayan, K. (Eds.), RDF123: from spreadsheets to RDF. Proceedings of the Semantic Web—ISWC 2008: 7th International Semantic Web Conference, ISWC 2008, Karlsruhe, Germany, October 26–30. Springer Berlin Heidelberg, Berlin, Heidelberg, pp. 451–466.

Huynh, D., Karger, D., 2009. Parallax and Companion: Set-Based Browsing for the Data Web. In: WWW Conference. ACM, vol. 61.

Kanehisa, M., Sato, Y., Kawashima, M., Furumichi, M., Tanabe, M., 2016. KEGG as a reference resource for gene and protein annotation. Nucleic Acids Res. 44, D457–D462.

Kobilarov, G., Dickinson, I., 2008. In: Humboldt: exploring linked data.LDOW.

Livingston, K.M., Bada, M., Baumgartner, W.A., Hunter, L.E., 2015. KaBOB: ontology-based semantic integration of biomedical databases. BMC Bioinformatics 16, 126.

Momtchev, V., Peychev, D., Primov, T., Georgiev, G., 2009. In: Expanding the pathway and interaction knowledge in linked life data.Proc. of International Semantic Web Challenge.

Philipp Heim, J.Z., Steffen, L., 2008. In: gFacet: a browser for the web of data.Proceedings of the International Workshop on Interacting With Multimedia Content in the Social Semantic Web (IMC-SSW'08).

Reck, R.P., 2003. Excel2rdf for Microsoft Windows. http://www.mindswap.org/rreck/~excel2rdf.shtml.

Russell, A., 2008. In: NITELIGHT: a graphical editor for SPARQL queries. Proceedings of the 2007 International Conference on Posters and Demonstrations. vol. 401. CEUR-WS.org, Karlsruhe, Germany.

Schweiger, D., Trajanoski, Z., Pabinger, S., 2014. SPARQLGraph: a web-based platform for graphically querying biological semantic web databases. BMC Bioinformatics 15, 279.

Smiley, D., Pugh, E., 2009. Solr 1.4 Enterprise Search Server. Packt Publishing Ltd., Birmingham.

The UniProt, C., 2015. UniProt: a hub for protein information. Nucleic Acids Res. 43, D204–D212.

Toyoda, K.N.a.T., 2012. In: BioLOD.Org: ontology based integration of biological linked open data.Proceedings of the 4th International Workshop on Semantic Web Applications and Tools for the Life Sciences.

Zaki, N., Tennakoon, C., 2017. BioCarian: search engine for exploratory searches in heterogeneous biological databases. BMC Bioinformatics 18, 435.

Zaki, N., Efimov, D., Berengueres, J., 2013. Protein complex detection using interaction reliability assessment and weighted clustering coefficient. BMC Bioinformatics 14, 163.

Zaki N., Tennakoon C. and Al Ashwal H., 2017. Knowledge graph construction and search for biological databases. In: Research and Innovation in Information Systems (ICRIIS), 2017 International Conference on, IEEE, pp. 1–6.

CHAPTER 8

Healthcare Decision-Making Support Based on the Application of Big Data to Electronic Medical Records: A Knowledge Management Cycle

Javier Carnicero*, David Rojas†
*Health Service of Navarra, Pamplona, Spain
†Sistemas Avanzados de Tecnología (SATEC), Madrid, Spain

1. INTRODUCTION

The progress of medicine and information technology (IT) has created a new scene in which patients are demanding progressively greater and higher-quality care as well as access to more information and significant participation in their own healthcare processes. In addition, the aging of the population is causing a continuous increment of chronic diseases, and budgetary constraints lead to the structural necessity of controlling the costs of health systems up to the point of requiring a substantial transformation of their current model to ensure their sustainability and continuity.

Because any health system must reconcile budgetary stability with the achievement of a series of healthcare goals, both individual and collective, new ways to increase the efficiency and effectiveness of healthcare organizations must be found. This can include the promotion of patient self-care, personalized medicine, or precision medicine, for example. One of the most important initiatives to achieve these goals is more ambitious knowledge management, based on data from corporate health information systems. Healthcare organizations are great generators and consumers of information at the same time, but nowadays they are very far from extracting all the potential value associated with these data. The application of Big Data solutions to this issue represents one of the best opportunities—as well as one of the biggest challenges—for the evolution of health systems.

2. KNOWLEDGE MANAGEMENT CYCLE IN HEALTHCARE

In every organization, the aim of knowledge generation is the continuous improvement of business processes, thus establishing a virtuous cycle. In a healthcare environment, the

Leveraging Biomedical and Healthcare Data
https://doi.org/10.1016/B978-0-12-809556-0.00008-3

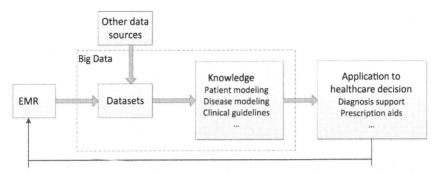

Fig. 1 Clinical knowledge management cycle. The EMR is an essential tool for providing healthcare as well as a source of knowledge that can be complemented with data provided by other sources. With the addition of features such as patient empowerment, personalized medicine, and clinical decision-making support systems, the EMR is also the keystone of a learning health system, for its data are used both to provide and improve healthcare.

clinical information systems and especially the electronic medical record (EMR) must be consolidated as the alpha and omega of a knowledge management cycle that leads to the improvement of medical care and the global performance of health systems.

The EMR has always been primarily an essential instrument for healthcare, but it is also a source of knowledge and, if necessary, it may even be used as a legal document. Therefore, EMR data have been processed for years to improve the quality of healthcare. This information has been used as a cornerstone for clinical research and education and is also crucial to enhance epidemiologic surveillance and detect public health situations that require an intervention, among other possible applications. In recent years, a patient's empowerment, personalized medicine, and clinical decision-making support systems have been added to the uses of EMR data. The final goal must be the creation of a health-care infrastructure that enables continuous learning and improvement, based on real-time knowledge production and application that allows health systems to be more efficient and effective in a predictive, preventive, personalized, and participatory way (Ross et al., 2014). This brings up the concept of a learning health system (LHS), envisioned as an integrated health system "in which progress in science, informatics, and care culture align to generate new knowledge as an ongoing, natural by-product of the care experience, and seamlessly refine and deliver best practices for continuous improvement in health and healthcare" (Friedman et al., 2015). Fig. 1 shows the basic scheme for the knowledge management cycle proposed in this article. Its main components are described below.

2.1 Electronic Medical Record

The EMR is a basic tool for providing healthcare, and thus essential to clinicians. By consulting it, they can access all the patient's information they need to know during a health-care episode as well as make clinical decisions and register all the new data generated by them. Thus, the EMR becomes the most valuable information repository of any health-care organization, and consequently a critical factor for the success of any Big Data-based solution to be applied as a contribution to the transformation of health systems.

In fact, the EMR may already be considered as a Big Data system itself because the information registered on the patients' records meets the requirements of volume, velocity, and variety traditionally attributed to the definition of Big Data (Laney, 2001). Furthermore, the Health Information Technology (HITECH) Act of 2009 in the United States included a series of incentives in order to encourage the adoption and certification of EMR systems, leading to an increase of almost 80%. Some calculations estimate that EMRs may have been used to document around 1 billion patient consultations per year (Ross et al., 2014).

It must be pointed out that the information generated during a medical consultation can be used not only for individual and collective healthcare, but also to optimize the management of the used resources, the invoicing of the delivered care services, the long-term results assessment, the strategic planning, and, as stated above, the clinical research or education (Rojas and Carnicero, 2015).

2.2 Other Data Sources

Other than the information stored in the EMRs, there is a huge amount of available complementary information related to diseases and other medical conditions, genetics (omics data such as genome sequences, RNA and microRNA, or proteomics), and drugs and therapies, among others (Ross et al., 2014). In addition, the improvement of connectivity and patient empowerment have stimulated the creation of new data sources, such as social networks, wearable devices, environmental sensors, or patient-reported outcomes (Weng and Kahn, 2016).

Furthermore, a health system implies the existence of a health cluster as a result of the interaction or collaboration of several public and private entities, such as central or federal governments, regional or local authorities, hospitals, primary care centers, health professionals, public health services, insurance companies and other healthcare financers, universities for the education and training of clinicians, research centers, patients' associations, pharmaceutical companies, and the health technology industry (Rojas and Carnicero, 2015).

2.3 Dataset Extraction

Obviously, not all the information will be useful in every specific case. For example, in the case of a clinical decision-making process related to a diabetic patient, the data of interest will be those related to patients who meet the same characteristics. As a consequence, a phenotyping procedure, very similar to the ones used for the selection of

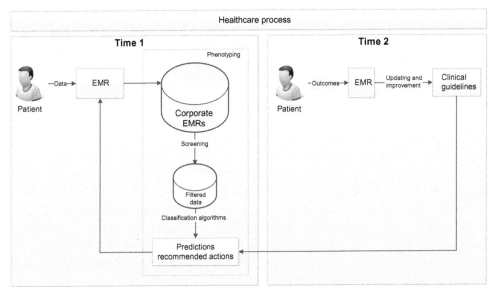

Fig. 2 Phenotyping, decision-making support, and learning. Not all the information present in the EMRs will be useful in every single scenario. The clinical knowledge management process must provide filter mechanisms for the selection of the data applicable and necessary in each case.

patient cohorts in clinical trials, must be applied. Fig. 2 shows the basic stages and features of phenotyping as well as the generation of knowledge in order to improve healthcare.

As explained above, a healthcare organization generates huge amounts of information, which translates into a great diversity of data, sources, structures, and uses. The clinical knowledge management process must provide filters for the selection of the data necessary in a determined case, which will depend on the information necessities to be solved in that specific situation. When information from several sources is managed, a data linkage procedure must be applied in order to integrate and harmonize these data. Data linkage consists of five steps (Bradley et al., 2010):

1. Identify the data sources that can be useful.
2. Obtain the necessary approvals, which may involve institutional ethics boards, regulatory authorities, and funding sources.
3. Select the variables that will be used to link the databases and individually clean the datasets.
4. Determine the best method and algorithms for linking the databases.
5. Evaluate the quality of the link established between data sources.

The use of advanced computing technologies will be required for the extraction of useful structured datasets. Natural language processing (NLP) must be noted as one of the most helpful tools in order to extract clinical data from narrative texts (Liao et al., 2015). Fig. 3 shows a conceptual scheme for data linkage applied to the development of clinical decision support systems.

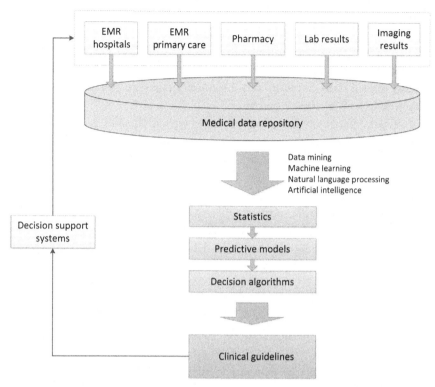

Fig. 3 Data linkage applied to decision-making support. The recollection and combination of data provided by EMRs and other health information sources as well as their processing by a set of advanced IT tools allow the development and evolution of clinical decision support systems.

2.4 Generation of Knowledge

The use of an EMR allows the clinicians to thoroughly register healthcare practices on structured datasets. This way, the EMR has become a source of evidence for the monitoring and assessment of clinical outcomes related to specific treatments and specific groups of patients. Although the paradigm of evidence-based medicine still prevails, the increasing use of EMRs in healthcare practice has brought up the concept of practice-based medicine. In this new approach, knowledge derived from the bibliographical databases can be complemented with empirical knowledge extracted on a real-time basis from the EMRs of selected patients (Martín-Sánchez and Verspoor, 2014).

In addition, the analysis of massive datasets depends on the efficiency of the analytics methodology. Traditional computing tools are not capable of managing very large and complex datasets, so new methods of distributed analytics are being developed. In other words, distributed processing allows "taking the analytics to the data" instead of the traditional "taking the data to the analytics." This new approach fits the distributed nature of patient information, which is stored in several systems, health centers, organizations, or even regions or countries. However, these distributed analytics techniques must ensure the compliance of data protection regulations through deidentification mechanisms to protect patient privacy (Weng and Kahn, 2016). Once the balance between data availability and confidentiality is guaranteed, the potential of EMR-based Big Data analysis will allow health systems to considerably improve the efficiency and effectiveness of clinical trials, reduce their duration and, in consequence, accelerate the incorporation of new treatments to the clinical practice.

The addition of patient-reported outcomes as well as changes in healthcare delivery and organization may also contribute to the development of the learning health system mentioned above, thus making the information and evidence generated in clinical practice available for research purposes. The launch of pragmatic clinical trials (PCTs), focused on developing clinical decision-making support systems by evaluating interventions in "real-world" healthcare settings, will also be a key element for the establishment of the LHS. Nevertheless, this will require new ethical guidelines to assess consent and risk, new methods to process observational data, and more effective partnerships of healthcare organizations and systems (Richesson et al., 2013).

2.5 Application to Healthcare Decision-Making Process

Health-related Big Data streams can be classified into three categories (Hansen et al., 2014):

1. Traditional healthcare data are generated within the health system and stored in datasets such as EMRs, medical imaging tests, and lab reports or pathology results, among others, as shown in Fig. 3. By analyzing this information, clinicians can have a better understanding of disease outcomes and their risk factors, and researchers are allowed to identify potential causes of adverse events. In other words, healthcare organizations can use these data to improve their clinical procedures and reduce errors. Moreover, EMR systems can register detailed metadata sets about their use, which are very useful for auditing purposes. For instance, this allows the detection of technology-related safety issues and concerns, such as user-device interaction problems, and can help prevent adverse events (Kuziemsky et al., 2014).

2. "Omics" data deal with large-scale datasets in the biological and molecular fields, such as genomics, microbiomics, or proteomics, among others, as mentioned above. The study of this data leads to deeper knowledge about how diseases behave, thus allowing the acceleration of the individualization of medical treatments. Personalized medicine aims to customize the diagnosis of a disease and the subsequent therapy by taking into account the individual patient's characteristics, for instance, matching the right drug with the right dosage to the right patient at the right time. Moreover, gene sequencing and the use of the subsequent genetic data in diagnosis and treatment will be essential to the future of personalized medicine, with actions such as the prescription of drugs based on genomic profiles of individual patients, known as pharmacogenomics.

Combinations of several types of data must also be taken into account. The concept of personalized medicine seeks to combine the patient's EMR and genomic data in order to support the clinical decision-making process, making it predictive, personalized, preventive, and participatory, an idea known as "P4 Medicine" (Panahiazar et al., 2014). This will require the integration of clinical information, mainly patient records, and biological data such as genome or protein sequences. These data are generated from different and heterogeneous sources and have very diverse formats, so a data linkage procedure will be required.

3. By analyzing data from social media, researchers learn how individuals or groups use the Internet, social media, apps, sensor devices, wearable devices, or any other tools, and use that to inform people about their health status and how to improve or preserve it. In addition, the inclusion of geospatial and environmental information may potentiate the ability to interpret gathered data and extract new knowledge.

3. KEY POINTS FOR AN EMR-BASED BIG DATA APPROACH TO CLINICAL KNOWLEDGE MANAGEMENT

3.1 Semantic Interoperability

Interoperability can be defined as the feature that allows two or more different systems to exchange data without human intervention. Semantic interoperability is the area of interoperability focused on the homogenous interpretation of the data exchanged, so that any system can process them ensuring that their value, context, and meaning are preserved. Thus, data standardization is enabled and the information can be understood and interpreted in the same way by all participants (Abad and Carnicero, 2012).

For semantic interoperability to be attained, the following aspects must be considered:
- *Contextual information*: agents involved, versions, administrative data, etc.
- *Structure of the clinical data*: detailed clinical models, archetypes, templates, etc. It is necessary to determine which information is to be exchanged and with what structure.
- *Terminologies*: exchanged information must be coded in order to ensure that its meaning is preserved through transmission.

3.2 Data Structure and Granularity

Data must be stored in a way that makes it easy to consult and process. In an EMR, information is progressively aggregated, so it has to be sorted according to a structure that allows this without affecting its use for healthcare and other purposes. This structure must be defined through a data model, which must establish an appropriate data granularity. This term, when referring to data storage, is defined as the scale or level of detail in a dataset.

Granularity is closely related to terminologies because these need to have enough granularity for the clinicians to accurately express information with the same level of accuracy as they would when using paper-based medical records. For instance, SNOMED-CT (Systematized Nomenclature of Medicine—Clinical Terms) has a very high granularity because it includes more than 300,000 terms from different health science domains. The advantage of having such a wide-ranging terminology is that it makes it possible to establish a reference terminology with which to map out several existing EMR solutions. However, excessive granularity can make data aggregation more difficult, thus preventing the making of statistical inferences or the application to decision-making processes (Abad and Carnicero, 2012).

3.3 Information Quality

As stated above, health information is generated in a massive, diverse, and quick way, so extracting high-quality and real data from these datasets becomes an important issue. Data quality can be defined as a combination of the dimensions or components featured in Table 1 (Cai and Zhu, 2015).

Table 1 Dimensions of data quality

Dimension	Definition
Accessibility	Difficulty level for users to obtain data
Timeliness	Time delay from data generation and acquisition to utilization
Authorization	Whether an individual or organization has the right to use the data
Credibility	Trustworthiness of a source or message, with three key factors: reliability of data sources, data normalization, and the time when the data are produced
Definition/ documentation	Data specification, which includes data name, definition, ranges of valid values, standard formats, and business rules, among others
Metadata	Description of different aspects of the datasets to reduce the problems caused by misunderstanding or inconsistencies
Accuracy	Result of data comparison to a known reference value, when possible
Consistency	Whether the logical relationship between correlated data is correct and complete
Integrity	All characteristics of the data must be correct: data have a complete structure, and data values are standardized according to a data model and/or data type
Completeness	The values of all components of a single datum are valid
Auditability	Auditors can fairly evaluate data accuracy and integrity within rational time and manpower limits during the data use phase
Fitness	The amount of accessed data used by users and the degree to which the data produced matches users' needs in the aspects of indicator definition, elements, classification, and others
Readability	Ability of data content to be correctly explained according to known or well defined terms, attributes, units, codes, abbreviations, or other information
Structure	Level of difficulty in transforming semistructured or unstructured data to structured data through technology

3.4 Patients Privacy: Legal Framework and Bioethics

The implementation of Big Data solutions and tools applied to healthcare requires addressing not only the organizational and technological issues detailed above, but also several legal and ethical issues. From a legal point of view, the first cause of conflict is data propriety. As explained in previous sections, data properly processed and analyzed can be turned into knowledge, which can be easily made profitable.

Tech giants have been working this angle for years. Google provides personalized advertisements based on the user's navigation and search history, and Facebook is using its users' data for sociological research purposes, having even tried to take possession of this information in a completely unilateral way. Because the generation, registry, and processing of such huge amounts of information require a powerful IT infrastructure, and therefore a large investment by these companies, their intention to make it profitable may be considered legitimate up to a certain point, especially when they are not charging users for the service provided. However, limitations regarding the use of the stored data must be clearly established, something that seems to be far from being solved through the current legal framework, which is quite confusing. For instance, in the case of Spain, this framework combines European Union, national, regional, and sectorial (both health and e-Government) regulations (Martínez and Rojas, 2014). Considering the current potential of Big Data solutions as well as the foreseeable one on the short and mid-term, a revision of these regulations seems to be most necessary.

Of course, this must be addressed with the goal of achieving a balance between the individual interests of patients (right to privacy) and professionals (legal certainty while performing healthcare and management duties) and the general interests of society (research, education, and improvement of healthcare services, among others). To that end, protocols must be defined that combine both a priori measures, such as data anonymization, and a posteriori measures, such as thorough audits regarding the data access and use. Regarding the human factor, one of the most crucial initiatives will always be raising the awareness of patients, professionals, and organizations (León, 2016).

From an ethical point of view, quite a few similarities to the legal field can be observed. The fact that IT is going to play an increasingly important role in health systems seems to be widely accepted, because its potential as a key instrument for the transformation of the current model is appreciated. Nevertheless, there is also great concern about the lack of transparency in the management of the large amounts of clinical data guarded by healthcare organizations. For this reason, the promotion of more and better control measures is backed by bioethics experts, starting with the development of a specific legal framework that can be translated into clear and visible actions, thus transmitting a sense of security to the public opinion and promoting trust in healthcare EMR-based Big Data solutions. Table 2 summarizes the main data sources, their major applications, and their requirements and limitations.

Table 2 Data sources, systems, requirements, and limitations

Data sources	Major applications	Requirements and limitations
Electronic medical record	– Healthcare quality improvement – Identify sources of adverse events – Clinical research	– Structure data model – Data quality – Data protection
Additional data about medical conditions, underlying genetics, medications, and treatment approaches	– Personalized medicine – Clinical decision support systems – Clinical research	– Ethical and legal issues – Data protection – Standardization of data – Data quality
Electronic medical record plus additional health data	– Pragmatic clinical trials – Clinical research – Patient empowerment	– Standardization – Data linkage – Data property
Health information system	– Performance assessment – Quality assessment – Planning of health services – Research and innovation	– Semantic interoperability – Data linkage

4. CONCLUSIONS

The need to transform healthcare services and health systems, their intensive use of data, and the adoption of personalized medicine, together with the availability of Big Data technologies, are leading to a new scenario of knowledge management in healthcare organizations. The main applications of Big Data to EMR are currently focused on patient safety, improvement of quality and efficiency of care, promotion of personalized medicine, patient empowerment, and clinical research. The key points for data extraction and analysis are semantic interoperability, a structured data model with an appropriate granularity, and information quality. Besides, legal and ethical issues must be considered in any Big Data project based on the use of EMRs and other health information sources in order to achieve a balance between the individual rights of patients and professionals and the general interests of society.

REFERENCES

Abad, I., Carnicero, J., 2012. International exchange of health information. In: Carnicero, J., Fernández, A. (Eds.), IX SEIS Report, "eHealth Handbook for Managers of Healthcare Services and Systems". Economic Commission for Latin American and the Caribbean (ECLAC) and Sociedad Española de Informática de la Salud (SEIS).

Bradley, C.J., Penberthy, L., Devers, K.J., Holden, D.J., 2010. Health services research and data linkages: issues, methods, and directions for the future. Health Serv. Res. 45 (5 Part II), 1468–1488.

Cai, L., Zhu, Y., 2015. The challenges of data quality and data quality assessment in the big data era. Data Sci. J. 14 (2), 1–10.

Friedman, C., et al., 2015. Toward a science of learning systems: a research agenda for the high-functioning learning health system. J. Am. Med. Inform. Assoc. 22, 43–50.

Hansen, M.M., Miron-Shatz, T., Lau, A.Y.S., Paton, C., 2014. Big data in science and healthcare: a review of recent literature and perspectives. Contribution of the IMIA Social Media Working Group. Yearb. Med. Inform. 9 (1), 21–26.

Kuziemsky, C.E., Monkman, H., Petersen, C., Weber, J., Borycki, E.M., Adams, S., Collins, S., 2014. Big data in healthcare—defining the digital persona through user contexts from the micro to the macro. Contribution of the IMIA Organizational and Social Issues WG. Yearb. Med. Inform. 9, 82–89.

Laney, D., 2001. 3D Data Management: Controlling Data Volume, Velocity, and Variety. META Group (Now Gartner).

León, P., 2016. Bioética y explotación de grandes conjuntos de datos. In: Carnicero, J., Rojas, D. (Eds.), La explotación de datos de salud: Retos, oportunidades y límites. Sociedad Española de Informática de la Salud (SEIS).

Liao, K.P., Cai, T., Savova, G.K., Murphy, S.N., Karlson, E.W., Ananthakrishnan, A.N., et al., 2015. Development of phenotype algorithms using electronic medical records and incorporating natural language processing. BMJ. https://doi.org/10.1136/bmj.h1885.

Martínez, R., Rojas, D., 2014. Gestión de la seguridad de la información en atención primaria y uso responsable de Internet y de las redes sociales. In: Carnicero, J., Fernández, A., Rojas, D. (Eds.), X SEIS Report, "Manual de Salud electrónica para directivos de servicios y sistemas de salud. Volumen II: Aplicaciones de las TIC a la atención primaria de salud". Economic Commission for Latin American and the Caribbean (ECLAC) and Sociedad Española de Informática de la Salud (SEIS).

Martín-Sánchez, F., Verspoor, K., 2014. Big data in medicine is driving big changes. Yearb. Med. Inform. 9, 14–20.

Panahiazar, M., Taslimitehrani, V., Jadhav, A., Pathak, J., 2014. In: Empowering personalized medicine with big data and semantic web technology: promises, challenges, and use cases.Proc. IEEE Int. Conf. Big. Data 2014, pp. 790–795.

Richesson, R.L., Hammond, W.E., Nahm, M., et al., 2013. Electronic health records based phenotyping in next-generation clinical trials: a perspective from the NIH Health Care Systems Collaboratory. J. Am. Med. Inform. Assoc. 20, e226–e231.

Rojas, D., Carnicero, J., 2015. A model of information system for healthcare: global vision and integrated data flows. Adv. Med. Biol. 82, 39–67.

Ross, M.K., Wei, W., Ohno-Machado, L., 2014. "Big data" and the electronic health record. Yearb. Med. Inform. 2014, 97–104.

Weng, C., Kahn, M.G., 2016. Clinical research informatics for big data and precision medicine. Yearb. Med. Inform. 2016 (1), 211–218.

CHAPTER 9

Computational Modeling in Global Infectious Disease Epidemiology

Ali Alawieh*, Zahraa Sabra[†,‡], Fadi A. Zaraket[‡]
*Department of Microbiology and Immunology, Medical University of South Carolina, Charleston, SC, United States
[†]Department of Neurosciences, Medical University of South Carolina, Charleston, SC, United States
[‡]Department of Electrical and Computer Engineering, Maroun Semaan faculty of engineering and architecture, American University of Beirut, Beirut, Lebanon

1. INTRODUCTION

Due to the epidemiological properties of infectious diseases (ID) in terms of contagious spread and dissemination at the population level, risk analysis and management strategies have been a significant component of epidemiological surveillance and population-level management of ID and the associated healthcare impact (Angulo et al., 2009; Duintjer Tebbens et al., 2008a). Prevention has been the cornerstone approach in the management of ID by a combination of active surveillance and mandatory reporting systems, immunization and exposure control, early treatment and containment of outbreaks, and adequate monitoring of the use of antimicrobials. Although these efforts have significantly reduced the burden of infectious diseases, especially in the developing world (Vos et al., 2015), rising bacterial antimicrobial resistance, misconceptions concerning vaccination, and war and conflict have recently increased the worldwide burden of ID (Munguia and Nizet, 2017; Omer et al., 2009; Sharara and Kanj, 2014; Alawieh et al., 2014; Bizri et al., 2014). These challenges have triggered increased interest in improving treatment options for ID, especially treatment-resistant bacteria, previously untreatable viral infections, and neglected pathogens while simultaneously motivating international-level and governmental-level surveillance programs to track outbreaks, evaluate risks, and assess the results of interventions.

Surveillance efforts have served an additional purpose of curating large volume datasets of disease progression, demographic and geographic patterns, and patterns of resistance to therapeutics as well as changes in these variables over time. These datasets provide a useful resource for analysis of trends, assessment of response to interventions at population levels, anticipation of upcoming outbreaks, and prognosis on change in resistance and spread patterns of microbial agents (Hay et al., 2013).

In this chapter, we review two approaches that we have recently developed to utilize data curated from surveillance programs to answer key questions in the field of infectious diseases. In the first section, we describe the use of machine learning to understand the

Leveraging Biomedical and Healthcare Data
https://doi.org/10.1016/B978-0-12-809556-0.00009-5

progression of bacterial resistance across Europe over a 10-year period using population-level data. We then use the abstracted patterns to predict future progression and automatically detect outbreaks and responses to interventions. In the second section, we describe an individual-based model to predict the risk of a poliovirus outbreak in Lebanon after outbreaks in nearby Syria and to determine the minimum requirements of population-level immunization coverage to prevent outbreaks.

2. PREDICTION OF PROGRESSION OF BACTERIAL RESISTANCE

Epidemiological surveillance datasets have monitored changes in the pattern of bacterial resistance to antimicrobials for several decades by national and international agencies. However, this data has been in the form of traditional epidemiological maps, curves, and changes in percentages over time (Rossolini and Mantengoli, 2008). With the advances of machine-learning algorithms and the availability of multinational datasets of bacterial resistance trends over time, we have recently described a new approach to monitor and predict bacterial resistance over time for surveillance and intervention purposes (Alawieh et al., 2015). The approach using bacterial and antimicrobial resistance distribution maps (BARDmaps) will be described in this section. BARDmaps were used to predict bacterial resistance progression specifically in Europe using a publicly available repository of bacterial resistance monitoring data (de Kraker and van de Sande-Bruinsma, 2007).

2.1 BARDmaps Task and Data Description

The overall aim behind BARDmaps is the prediction of the level of resistance for a specific bacterium against an antimicrobial agent for upcoming years based on historical trends. Simultaneously, BARDmaps will also provide a platform for the automated detection of outbreaks and monitoring similarity in trends between different bacteria-antimicrobial combinations. The data used to train, run, and execute BARDmaps were obtained from the European Antibiotic Resistance Surveillance Network (ERAS-Net). Data in ERAS-Net is provided as the percentage of sampled isolates of a bacterial species that are resistant to a specific antimicrobial agent at a national level using antimicrobial susceptibility tests performed at central standardized labs. The data included seven bacterial species and 29 antimicrobial agents between the years 1999 and 2012 for 30 European countries (Alawieh et al., 2015).

2.2 Design of BARDmaps

The predictions and analysis of trends in BARDmaps were based on using hidden Markov models (HMM) with expectation maximization (EM) (Alawieh et al., 2015). HMMs are built on the assumption that the data can be modeled using a Markov process in which hidden states control a set of observations. The observations, which are in this case the changes in resistance status of a certain bacterium to an antimicrobial, are then used to derive these hidden states. Two components were included in BARDmaps that serve two different aspects of the tool: a structural component and a behavioral (predictive component). The structural component includes detection of resistance change patterns, comparison of patterns, and visualization of the trends. The patterns detected in the structural component will be used as input to the behavioral model to predict future trends.

In detail, the following sequence describes the different steps of the process implemented in BARDmaps.[1]

2.2.1 Initial Processing

Data were curated from ERAS-Net as the percentage of resistant isolates for each bacteria-antimicrobial-country combination across the available years. Bacteria-antimicrobial pairs (BAPs) were defined by a bacteria β tested against an antimicrobial δ where the pair (β, δ) has a unique value per country per year. Differences for each BAP per country were independently calculated across intervals of 1, 2, 3, 4, and 5 years, and each of these values for each BAP was used as an input to the structural model in BARDmaps. The maps representing the BARDmaps are then visualized as a network of nodes where each node represents the resistance percentage at a certain year, and each edge represents the difference between any two consecutive or nonconsecutive years. The differences computed in the BARDmaps were used as observations to train an HMM machine.

2.2.2 Structural Analyses

Following the derivation of BARDmaps, the detection of abrupt changes of bacterial resistance patterns was pursued to detect a sudden increase in resistance (representing outbreaks) or a sudden decrease in resistance (representing interventions). To automatically detect these changes, we computed, for each BAP, the moving of the average of the change in resistance between any two consecutive years, and an abrupt change was defined as a change that is two standard deviations above the moving average for a given BAP. The second component of structural analysis includes the comparison of trends among different BAPs to detect which BAPs perform similarly over time. We performed a topological sorting of BARDmaps by computing trend vectors that represent the sequence of change over time for each BAP. We then calculated the distance metrics between any two specific vectors to detect similarities among the BAPs, as shown in Alawieh et al. (2015).

---

[1] Please refer to the following link for detailed source code implemented in MATLAB to run BARDmaps on the available dataset: https://github.com/alialawieh1/BARDmaps.

2.2.3 Behavioral (Predictive) Analyses

For predicting future trends in resistance, we used the BARDmaps as inputs to an HMM model.[1] The observed differences in resistance patterns over time for each BAP were used as inputs, and the output of the HMM was set as a predicted score for different candidate resistance levels. To illustrate, a set of predefined resistance levels at 5% intervals (0%, 5%, 10%, etc.) or 2% intervals and the output of the HMM is a score for each level. Then, a weighted average by score of the different resistance levels was used to compute the predicted resistance at 1-year or 5-year intervals. Predictions were then validated using actual resistance levels at predicted years.

2.3 Outcome and Performance of BARDmaps

Structural modeling implemented in BARDmaps was able to automatically detect outbreaks and interventions in multiple countries. For instance, two outbreaks of vancomycin-resistant *E. faecium* were detected in Germany from 2004 to 2007. Another example is the detection of a rapid increase in the proportion of penicillin-resistant *S. pneumoniae* in Spain in 2011 followed by a rapid decline, which corresponded to an outbreak in the country that responded well to the introduction of the 7-valent pneumococcal conjugate vaccine (Alawieh et al., 2015). Using trend analysis, BARDmaps was also able to sort bacterial pathogens by their similarity in resistance patterns. An interesting finding was the similar pattern of progression of carbapenem resistance in *E. coli* and vancomycin resistance in *E. faecium*, where both carbapenems and vancomycin are the drugs of last resort for each of the bacteria. This is likely indicative that the increased exposure to these drugs is potentially driving the progression, given the absence of similar genetic and/or biochemical similarities between the two BAPs. Finally, we have demonstrated that HMM models can provide a high prediction reliability where predicted and actual values are highly correlated with a Pearson's coefficient $R^2 = 0.86–0.98$ and a slope of 0.93–0.99, indicating near identity between the two variables (Alawieh et al., 2015). We have also compared predictions in BARDmaps to those of linear regression models and have shown the superiority of BARDmaps with a correlation coefficient between predictions and real values of $R^2 = 0.97$ at a 2-year prediction compared to $R^2 = 0.88$ with the best regression model (Alawieh et al., 2015).

2.4 Future Directions

BARDmaps provides one example on how machine-learning tools can be exploited to mine and utilize big datasets in the field of population epidemiology in infectious diseases. Subsequent work will still benefit from developing real-time prediction models that are updated automatically with new data provided in public repositories to provide optimized predictions at a regular basis. In addition, moving the application of this approach to the international level may provide a platform to track the origins of outbreaks internationally, guide antimicrobial stewardship programs, and optimize efforts to control outbreaks. Finally, a similar approach can be used at a smaller scale to study geographical patterns within countries and across hospitals to determine priorities for interventions and to detect outbreaks early.

3. POLIOVIRUS PREDICTION

Using a different paradigm, this section discusses the use of an individual-based computational model to predict outbreak progression in a confined geographical and demographic population as well as a response to intervention. The example, used here as a platform to describe the utility of an individual-based model, is the prediction of a potential poliovirus outbreak in Lebanon, a developing country, after a new outbreak in its neighboring Syria, and how the prediction guides the efforts of the national immunization campaign to prevent an outbreak.

3.1 Model Aims and Data Inputs

The wild-type poliovirus (WPV) infection has been eradicated from the majority of countries globally; however, local foci of WPV transmission in Pakistan, Nigeria, and Afghanistan are still present and lead to recurrent outbreaks of WPV in the Middle East or Africa during periods of instability and uncontrolled population migration (Abimbola et al., 2013; Alawieh et al., 2017). This provides a threat to geographically adjacent WPV-free countries. An example is a recent WPV outbreak in Syria during the ongoing civil war that began in 2013. The outbreak brought concern about similar outbreaks happening in neighboring countries such as Lebanon and triggered international efforts to contain these potential outbreaks by massive immunization campaigns (Akil and Ahmad, 2016). Facing this threat, the Lebanese government planned a massive WPV immunization campaign among the Lebanese population and the incoming refugees (Alawieh et al., 2017). Although the threat of an outbreak is present, there was no clear strategy to quantitate the extent of this threat and to evaluate the needs that the campaign should aim to achieve to ensure the low risk of a new outbreak in Lebanon. Therefore, we developed an individual-based model to predict how WPV may progress in Lebanon after one or a few cases of WPV infection infiltrated the country before the immunization campaign as well as when different levels of national immunization coverage are achieved.

To design a model that can simulate disease progression at the individual level, we curated multiple public and disease-related datasets. First, to simulate disease process, we mined the scientific literature for the infectivity, incubation time, duration of infection, likelihood of clinical manifestation, and mortality in WPV cases by different age groups (children and adults). Then, we curated demographic data about the different resident and immigrant populations in Lebanon (Lebanese, Syrian, and Palestinian) at the household level as well as their geographic distribution across Lebanese provinces using the datasets published by the Lebanese government, the United Nations Development Program (UNDP), and the United Nations High Commissioner for Refugees (UNHCR). Finally, we retrieved the data on immunization coverage in different populations and geographic locations in Lebanon using the Lebanese Ministry of Public Health databases. These inputs were then used to simulate the Lebanese scenario in case WPV-infected individuals make it across the border among Syrian refugees (Alawieh et al., 2017).

3.2 Model Design

The overall design of the model consists of three steps:

- Generating a simulated geographically distributed set of individuals belonging to households within the different Lebanese provinces, and simultaneously distributed over nationality, age groups, and immunization status.
- Simulating disease progression if one or multiple WPV-infected individuals cross the border based on individual-individual contacts.
- Identifying the minimum national vaccination coverage required to prevent a potential outbreak.

The following assumptions were made before initiating the model: 1. contact between infected and nonimmunized individuals results in infection, 2. contact between an infected and an immunized individual does not result in infection, and 3. individuals within the same household have continuous direct contact. Further, each individual based on nationality is assigned a specific (random within a predefined range) number of daily contacts at the household, province, and national levels. The model source code and detailed description of runs are provided.[2]

3.2.1 Simulating the Lebanese Demographics

To construct the Lebanese demographics, we first started by defining an individual by a set of features that include the home site or province, the subpopulation (Lebanese, Syrian, Palestinian), the household size, the household, the individual number in the household, the immunization status (fully, partially, or not immune), the age group, and whether he/she has direct access to medical care. The distribution of households obtained from sources described above was used to populate a hierarchical tree for each subpopulation (Lebanese, Syrian, or Palestinian), with the root being the country followed by the geographical site, the subpopulation, and the number of individuals within the household. Following initiation of household identities and features, the individual features, described above, were then assigned semirandomly where general roles in the input datasets have been honored while allowing for random allocation of individuals. This process

---

[2] Please refer to the following link for detailed source code implemented in MATLAB to run the polio prediction model: https://github.com/alialawieh1/PolioModel.

was repeated several times for validation of findings as described below. Following the allocation of individual features, the outcome of the population generation algorithm is a set of vectors containing the demographic, geographic, and immune features of each individual (Alawieh et al., 2017).

3.2.2 Propagating a Potential Outbreak

To simulate a potential outbreak, we defined one or more cases of WPV-infected individuals introduced into one of the different Lebanese provinces. Each case was defined using a set of additional disease-related variables that includes duration of infection, latency period (time between exposure and infectivity), latency period (time between infectivity and symptomatic manifestation), symptomatic status, and infectivity. These variables were derived from published literature, often reported as a mean/median and range. Therefore, a skewed Gaussian distribution was used to generate random values within the published constraints for each variable.

Following the definition of case(s), a Monte Carlo-based simulation was then used to simulate a potential outbreak using the assumptions above. In addition, we added a set of rules:

- If an individual is infected, the case may or may not be symptomatic.
- If an individual is symptomatic (5% of cases) and has access to medical care, he/she will be detected and isolated.
- If an infected individual develops paralysis (0.5% of cases), the individual will not affect others.
- The simulation is run on a day-to-day basis.

These simulations were run for 1000–10,000 runs while reinitiating all randomly assigned variables to ensure the robustness of the model, and the progression of the number of cases was calculated within a certain duration of time, or until one or a few cases are detected. Comparison of how a potential outbreak propagates (duration to start, rate of evolution, number of cases, and time to be detected) was performed when the number of initial WPV cases was changed, the location of the initial cases was changed, and the nationality of the individual was changed to determine the high-risk scenarios.

3.2.3 Determining the Threshold for Vaccination Coverage

The simulations above were then reexecuted while replacing the national immunization coverage before the national vaccination campaign (∼95% in adults and ∼71% in children) with that at different levels of national coverage, including the levels achieved by the immunization campaign (97.8%). The propagation of outbreaks was then performed as described above.

3.3 Model Results

Using the above model, we demonstrated an alarming threat to Lebanon if actions were not taken where it was anticipated that at least 1000 cases of WPV-infected individuals will occur in the country 30 days after one case is introduced from Syria, regardless of which province was the host. Provinces near the Syrian border were at significantly higher risk given the higher density of Syrian immigrants. At this same baseline immunization coverage, it was anticipated that within 12 days of an introduced case, the number of infected individuals will start doubling on a daily basis, leading to 15 deaths and 30 paralyzed individuals within the first month. However, following the national immunization campaign and a 97.8% coverage with vaccination, this threat was significantly reduced where the maximum number of infected individuals after the introduction of a single case to the country was five, and the probability of an outbreak was reduced by 15-fold. By also simulating outbreak progression at different national immunization coverage, we determined that the risk of potential outbreaks exponentially increases when the national immunization coverage drops below 90%, including all subpopulations, and is nearly nonexistent when the coverage is above 97% (Alawieh et al., 2017).

3.4 Future Directions

The findings described in this model were further supported by actual events as today, 5 years after the start of the outbreak in Syria, and with maintenance of high national immunization coverage in Lebanon, there has been no single case of WPV infection reported in the country. This model depicted first for the poliovirus infection, given its high national and international burden and potentially fatal outcome in children, can be easily translated to other models of infectious diseases, including rare as well as more common pathogens. It can also be supplemented by a reverse approach where an outbreak can be decoded to determine the origin and spread after it has already occurred to determine areas that need prioritized intervention.

4. CONCLUSIONS

In this chapter, we have described two novel applications of computational modeling in the field of infectious disease epidemiology that study disease-specific epidemiology at international or national levels. In addition to the approaches described here, there has been significant advances in the application of computational modeling in infectious disease epidemiology over the past decade that included, among others, the study of spatial transmission of disease and population migration (e.g., GLEaM model (Balcan et al., 2010), and influenza propagation in gatherings (Shi et al., 2010)), the study of disease transmission in hospital settings (Cooper and Lipsitch, 2004; Donker et al., 2010), the study of the impact of vaccination (Burke et al., 2006; Duintjer Tebbens et al., 2008b), and the study of pathophysiological features impacting infectious disease progression (Castiglione et al., 2004).

REFERENCES

Abimbola, S., Malik, A.U., Mansoor, G.F., 2013. The final push for polio eradication: addressing the challenge of violence in Afghanistan, Pakistan, and Nigeria. PLoS Med. 10(10). e1001529.

Akil, L., Ahmad, H.A., 2016. The recent outbreaks and reemergence of poliovirus in war and conflict-affected areas. Int. J. Infect. Dis. 49, 40–46.

Alawieh, A., Musharrafieh, U., Jaber, A., Berry, A., Ghosn, N., Bizri, A.R., 2014. Revisiting leishmaniasis in the time of war: the Syrian conflict and the Lebanese outbreak. Int. J. Infect. Dis. 29, 115–119.

Alawieh, A., Sabra, Z., Bizri, A.R., Davies, C., White, R., Zaraket, F.A., 2015. A computational model to monitor and predict trends in bacterial resistance. J. Glob. Antimicrob. Resist. 3 (3), 174–183.

Alawieh, A., Sabra, Z., Langley, E.F., Bizri, A.R., Hamadeh, R., Zaraket, F.A., 2017. Assessing the impact of the Lebanese National Polio Immunization Campaign using a population-based computational model. BMC Public Health 17 (1), 902.

Angulo, F.J., Collignon, P., Powers, J.H., Chiller, T.M., Aidara-Kane, A., Aarestrup, F.M., 2009. World Health Organization ranking of antimicrobials according to their importance in human medicine: a critical step for developing risk management strategies for the use of antimicrobials in food production animals. Clin. Infect. Dis. 49 (1), 132–141.

Balcan, D., Goncalves, B., Hu, H., Ramasco, J.J., Colizza, V., Vespignani, A., 2010. Modeling the spatial spread of infectious diseases: the GLobal Epidemic and Mobility computational model. J. Comput. Sci. 1 (3), 132–145.

Bizri, A., Alawieh, A., Ghosn, N., Berry, A., Musharrafieh, U., 2014. Challenges facing human rabies control: the Lebanese experience. Epidemiol. Infect. 142 (7), 1486–1494.

Burke, D.S., Epstein, J.M., Cummings, D.A., Parker, J.I., Cline, K.C., Singa, R.M., et al., 2006. Individual-based computational modeling of smallpox epidemic control strategies. Acad. Emerg. Med. 13 (11), 1142–1149.

Castiglione, F., Poccia, F., D'Offizi, G., Bernaschi, M., 2004. Mutation, fitness, viral diversity, and predictive markers of disease progression in a computational model of HIV type 1 infection. AIDS Res. Hum. Retroviruses 20 (12), 1314–1323.

Cooper, B., Lipsitch, M., 2004. The analysis of hospital infection data using hidden Markov models. Biostatistics 5 (2), 223–237.

de Kraker, M., van de Sande-Bruinsma, N., 2007. Trends in antimicrobial resistance in Europe: update of EARSS results. Euro Surveill. 12 (3), E070315 3.

Donker, T., Wallinga, J., Grundmann, H., 2010. Patient referral patterns and the spread of hospital-acquired infections through national health care networks. PLoS Comput. Biol. 6(3). e1000715.

Duintjer Tebbens, R.J., Pallansch, M.A., Kew, O.M., Sutter, R.W., Bruce Aylward, R., Watkins, M., et al., 2008a. Uncertainty and sensitivity analyses of a decision analytic model for posteradication polio risk management. Risk Anal. 28 (4), 855–876.

Duintjer Tebbens, R.J., Thompson, K.M., Hunink, M.G., Mazzuchi, T.A., Lewandowski, D., Kurowicka, D., et al., 2008b. Uncertainty and sensitivity analyses of a dynamic economic evaluation model for vaccination programs. Med. Decis. Making 28 (2), 182–200.

Hay, S.I., George, D.B., Moyes, C.L., Brownstein, J.S., 2013. Big data opportunities for global infectious disease surveillance. PLoS Med. 10(4). e1001413.

Munguia, J., Nizet, V., 2017. Pharmacological targeting of the host–pathogen interaction: alternatives to classical antibiotics to combat drug-resistant superbugs. Trends Pharmacol. Sci. 38 (5), 473–488.

Omer, S.B., Salmon, D.A., Orenstein, W.A., deHart, M.P., Halsey, N., 2009. Vaccine refusal, mandatory immunization, and the risks of vaccine-preventable diseases. N. Engl. J. Med. 360 (19), 1981–1988.

Rossolini, G.M., Mantengoli, E., 2008. Antimicrobial resistance in Europe and its potential impact on empirical therapy. Clin. Microbiol. Infect. 14 (Suppl 6), 2–8.

Sharara, S.L., Kanj, S.S., 2014. War and infectious diseases: challenges of the Syrian civil war. PLoS Pathog. 10(10). e1004438.

Shi, P., Keskinocak, P., Swann, J.L., Lee, B.Y., 2010. The impact of mass gatherings and holiday traveling on the course of an influenza pandemic: a computational model. BMC Public Health 10, 778.

Vos, T., Barber, R.M., Bell, B., Bertozzi-Villa, A., Biryukov, S., Bolliger, I., et al., 2015. Global, regional, and national incidence, prevalence, and years lived with disability for 301 acute and chronic diseases and injuries in 188 countries, 1990–2013: a systematic analysis for the global burden of disease study 2013. Lancet 386 (9995), 743–800.

CHAPTER 10

Semiautomatic Annotator for Medical NLP Applications: About the Tool

Mohamed Sabra*, Ali Alawieh†
*Department of Microbiology and Immunology, Medical University of South Carolina, Charleston, SC, United States
†Department of Electrical and Computer Engineering, American University of Beirut, Beirut, Lebanon

1. INTRODUCTION

With the expansion of scientific and social media, a wealth of information has accumulated as free text within online resources, including articles, studies, and social blogs. Mining, standardization, and extraction of this information would bring about new approaches for the analysis and discovery of knowledge within large text corpora specific to certain disciplines. Natural Language Processing (NLP) tasks deal with text corpora and map these corpora to output domains. Those output domains differ according to the NLP task. Example output domains are (1) another language in case of a machine translation task, (2) word classification in case of a part of speech annotation task, and (3) categories in an information extraction (IE) task.

There are two NLP approaches: supervised and unsupervised. Both approaches need a reference text corpus annotated with correct data alongside a testing text corpus to evaluate the accuracy of the NLP technique. A supervised approach will also require a training text corpus that is annotated in a manual or standard manner to learn a computational model (Zaraket and Jaber, 2013).

There is ongoing interest in applying NLP information and relational extraction tasks to biomedical literature and texts. Several tools have been developed in the field of medical and biological NLP to handle different types of corpora (Nédellec et al., 2013) including:
— Electronic medical records (EMR).
— Medical literature databases such as PubMed and Scopus.
— Genia event extraction (GE).
— Cancer genetics (CG).
— Pathway curation (PC).
— Gene regulation ontology (GRO).
— Gene regulation network in bacteria (GRN).
— Bacteria biotopes (BB).

Leveraging Biomedical and Healthcare Data
https://doi.org/10.1016/B978-0-12-809556-0.00010-1

Several annotators such as Knowtator (Ogren, 2006), Brat (Stenetorp et al., 2012), WordFreak (Morton and Lacivita, 2003), MMAX2 (Müller and Strube, 2006), Callisto (Day et al., 2004), and Turku Event Extraction System (TEES) (Björne and Salakoski, 2013) are already present and were utilized in some biomedical applications such as the use of TEES in the curation of the gene regulatory network in bacteria (Nédellec et al., 2013). However, these annotators may not be optimal to all tasks of biomedical NLP, and optimized models of these annotators would better fit specific biomedical NLP tasks.

We are particularly interested in devising new biomedical NLP tools to convert the massive literature on complex diseases and conditions into a resource that fosters a better understanding of the disease processes and therapy. The biochemical basis of disease can be reduced into disturbances in gene products (such as proteins) that would lead to abnormal cellular biology, leading to pathology. Proteins and their relation to diseases have become important in recent medical research, especially that proteins are the final biochemical effectors and the most interesting targets for therapeutic intervention.

In the field of protein to disease relations, there are a huge number of articles that report protein changes during or after a certain disease. A wide range of proteins is reported by these studies, depending on the author's interest, the particular aspect of the studied disease, and the studied species model. However, the collective contribution of the entire spectrum of protein changes is the determinant of overall disease pathology. This spectrum of protein changes cannot be characterized by a few studies and requires cumulative evidence over decades of scientific research. Therefore, efforts to extract and curate the spectrum of protein changes from the large number of scientific reports about a given disease allow for both global and detailed analysis of the molecular architecture of diseases. The analysis features the full network of protein interactions responsible for disease phenotype and symptoms.

The aim of this work is to use semiautomatic annotation and biomedical NLP algorithms to provide a model for the curation and construction of a molecular architecture of pathological and disease conditions. The resulting model is particularly significant in the context of complex diseases where a large number of proteins contribute to the disease processes, such as Alzheimer's disease, stroke, myocardial infarct, and cancer metastasis.

Articles reporting protein changes after disease are stored in online biomedical databases such as PubMed and SCOPUS. The databases allow using a special API for the retrieval of the data in a predefined format such as EXtensible Markup Language (XML) and comma-separated values (CSV). The API takes a search key specific to the disease or disease process of interest as an input and returns the resulting articles that match the key in XML or CSV format. Each article presents evidence of the role of one or more proteins in activating a pathway of one or more diseases. To be able to treat a disease, medical researchers are interested in looking at the comprehensive picture.

A search in PubMed and Scopus returns 83,101 articles related to strokes. This magnitude of literature is hard to analyze manually and automated NLP tools that help in the selection of the relevant articles are needed.

Existing annotators fall short of performing this task because of the following reasons:

1. Most of the existing annotators need a professional software developer to deploy and run them, which is hard for a protein or disease specialist.
2. Some of those annotators could not handle large file sizes such as 100 MB while the data about brain ischemia is 688 MB.
3. They cannot parse XML and CSV files specific to biomedical indexing databases.
4. They do not generate needed statistics from the annotated data (such as frequency and cooccurrence of the annotated terms).
5. They are single document-based, and cross-document analysis is the target of this work.

Therefore, we need an annotator that allows for cross-document annotation, surpasses all the above shortcomings, and has the following abilities:

1. It should be user friendly and easy to install and use.
2. It should parse large files and extract important data from files such as titles, abstracts, IDs, and dates.
3. It should be able to display documentation aggregated across all articles and files, and allow for inspecting each one alone with its important components.
4. It should allow the intervention of the specialist to edit and redo the automatic annotations, including adding or deleting an annotation.
5. It should be able to create statistical metrics that describe the annotations and their occurrences while performing the NLP task.
6. It should be able to automatically suggest potential hits within biomedical texts with high accuracy.

That is why we built a Semiautomatic Annotator for Medical NLP Applications (SAMNA), which allows the user to load a large amount of data, including abstracts and titles of published articles that discuss a specific disease of interest. In addition to the required functionalities, SAMNA provides the following:

1. SAMNA visualizes the components of articles such as title, abstract, identifier tags, and date with color-sensitive annotations that highlight occurrences of terms of interest to the NLP task in the text.
2. SAMNA takes expert rules from specialists that specify features of the target annotations. It uses these rules to annotate additional terms not defined previously in the database. For example, rule "$\forall p \in P$: list of proteins; $p[1-9][1-9]^*$ is a protein" matches XYZ3 as a protein in case XYZ was a protein in database P.
3. SAMNA implements a distributional similarity algorithm using DISCO (Kolb, 2009) to find new annotations.

4. SAMNA saves a result file holding the important statistics of annotation occurrences and frequencies for each annotation. In this way, SAMNA provides a full protein spectrum of diseases for further analysis by biomedical investigators.

SAMNA was evaluated and utilized to study the following diseases:

Stroke (brain ischemia): We used SAMNA to construct, annotate, and curate a brain ischemia metaproteome that reflects the spectrum of protein changes after brain infarct. We used systems biology databases including UniProt (The UniProt Consortium, 2010) for protein accession retrieval and STRING (Szklarczyk et al., 2010) for protein interaction data to construct the interaction network among the extracted proteins. Curation of the proteome allowed for dissecting the role of different biochemical pathways in the amplification of injury after stroke and revealed, with the help of graph theory algorithms, the presence of a rich-club organization in the brain ischemia intractome. The detected pathways and the rich-club provide information on the optimal approaches to target injury mechanisms after stroke (Alawieh et al., 2015a).

Spinal cord injury (SCI): We also used SAMNA to curate the proteome of spinal cord injury (SCI) and we were able to characterize the full molecular architecture of SCI. Analysis of the curated proteome using graph theory revealed a modular organization of disease process that is centered around a core of cell survival decision proteins. Those core proteins provide the best targets for therapy. (Alawieh et al., 2015b)

Cancer metastasis: Using the multilabel capability of SAMNA, we applied the same approach to study epithelial-to-mesenchymal (EMT) transition, the process by which cancer cells develop migrating and metastatic potential. SAMNA allowed annotating for proteins, microRNAs, cancer subtypes, and drugs in order to detect the protein changes behind EMT and their relation to the cancer subtype, microRNA expression, and anticancer drugs.

2. THE SEMIAUTOMATIC ANNOTATOR

The semiautomatic annotator for medical NLP applications, SAMNA, allows the specialist to load XML, CSV, or text files. Those files can be extracted from PubMed and Scopus, for example. Then SAMNA allows semiautomatic annotation of these files.

2.1 SAMNA File Extraction

SAMNA loads the files and extracts from them the following needed components upon the request of the user.

(a) The title.
(b) The abstract.
(c) The ID (PMID for example), which is a unique ID that identifies the publication.
(d) The date of publication.

SAMNA can extract other parts of the articles in case scholars were interested in more details, such as the results, the figures, or the tables of the articles. Typically, the title and abstract are enough for annotation because they are manageable and easy for screening. They have the most important information about the publication work and a brief description of the results. Furthermore, while some papers are expensive, not open access, or not fully digitized yet, almost all papers have digitally accessible abstracts.

2.1.1 PubMed XML Structure

A search in the PubMed database results in an XML file with the following structure and tags. In the XML file, each article is embodied inside a node named PubMed. Article that contains the following tags.

(a) Article Title is the node holding the title of an article.
(b) Abstract is the node enclosing the abstract of an article.
(c) PMID is the node containing the PMID of the article.
(d) Date Completed is the node having the date of the article.

Several other details exist in the XML file that SAMNA ignores for now, such as journal details, authors names, and comments.

2.1.2 Scopus CSV and TXT Structure

CSV files extracted from Scopus are organized by column names. Each row represents an article. SAMNA is interested in

(a) The title column.
(b) The abstract column.
(c) The id column.
(d) The publication year column.

Several details exist in the CSV file that SAMNA ignores for now, such as authors names column, article link column, and source column.

2.2 SAMNA Navigation

SAMNA is a user-friendly annotator tool with a smooth article and annotation navigation process, a cashing process, and semiautomatic annotation algorithms.

(a) SAMNA allows the specialist to go between extracted articles one by one, forward and backward. It offers the choice to do that by clicking the next and previous button or through shortcut keyboard navigation keys.

(b) SAMNA applies the expert rules and distributional similarity annotation algorithm for each article and allows the user to validate the automatically found annotations.

(c) SAMNA allows the specialist to go to a specific article using an article identifier number. This gives the specialist the ability to continue working after a break.

(d) SAMNA learns while a specialist is performing manual annotation, and after a while the specialist can request to annotate all abstracts automatically while extracting statistics about annotation occurrences and frequencies. The specialist can then go over the automatic annotations and validate them. SAMNA produces typical accuracy measures to showcase the quality of the annotations.

(e) SAMNA implements shortcut keyboard keys for every important action. Thus, the specialist can annotate the articles in a faster way using the keyboard only. For example, the specialist can start highlighting the article while reading, and when she finds a term that should be annotated, she clicks CTRL+N where N is the number of words she wants to annotate starting from the cursor. SAMNA will annotate the number of specified words using the label of choice.

2.3 SAMNA Data Model

An annotation relates a term that occurs in text to a label that the specialist using SAMNA defines. SAMNA has a cashing system to load already defined lists of annotation labels and their associated terms. SAMNA stores new annotations and labels and saves statistics using spreadsheet files.

(a) SAMNA can work on several projects at once. Each project is organized in a workspace. SAMNA prompts the user for the workspace at first. The workspace may contain the annotation label and term lists.

(b) SAMNA saves the annotations in a result database. The specialist can have some statistics about each annotation and its frequency of occurrence. Each row in the result file is an annotation. The row contains the annotation term, the id of the article, the label it belongs too, the origin term from which the annotated term is deduced, the date of the article, the section where the annotation was found, and the position of the annotation in the text.

(c) SAMNA saves annotation labels and terms in a spreadsheet format. Those spreadsheets are used in the annotation process. It also saves editing actions in a spreadsheet format so the action will be memorized.

2.4 SAMNA Annotation

SAMNA has a manual annotation process, a direct string matching annotation process, and two automatic annotation processes that suggest new annotation.

(a) SAMNA annotates words based on a database of labels that is defined by the specialist. The specialist defines the label and associates it with a set of terms. The terms can be constructed automatically by the tool as well. Each label is also associated with a visualization legend. SAMNA uses the terms and legends to show the annotations in the text of the articles. For example, a label can be "protein," the terms can be the protein different names, and the legend can be a background color and an italic font.

(b) The specialist can also associate the label with rules. The rules are regular expressions that allow variants of the terms to match the annotation label.

(c) SAMNA highlights words in the article with exact matching to the terms and the rules and allows the annotator to validate the matches.

(d) SAMNA applies a distributional similarity algorithm based on the PubMed corpora from DISCO (Kolb, 2009) to find and suggest new annotations. The specialist also can validate the matches of the resulting annotations.

2.5 SAMNA User Editing

SAMNA allows the user to edit annotation labels, their associated annotation terms, their associated annotation rules, and the resulting annotations. SAMNA also allows the user to exclude a term or an article from the annotation process.

(a) SAMNA enables the specialist to add or remove an annotation label. To add a label, the specialist has to choose the label name and the associated legend. When the specialist adds a new label, a database entry of this label will be created and the specialist will be able to add and remove annotations to the label entry.

(b) SAMNA enables the specialist to add and remove new annotations to the label database. When adding (or removing) an annotation, the annotation will be highlighted (or unhighlighted) in all other articles. Any addition (or deletion) of an annotation will affect the label database.

(c) SAMNA enables the specialist to exclude a term from annotations in a specific article.

(d) SAMNA enables the specialist to exclude an article from the annotation process. This article will not be shown in the results. The user can specify part of the text as the reason for the exclusion.

(e) SAMNA allows the specialist to choose whether to apply the rules while searching for annotations in text.

(f) SAMNA gives the ability to undo actions.

(g) SAMNA offers shortcuts to do the important actions mentioned above.

2.5.1 SAMNA Analysis

SAMNA provides the results of the annotation in spreadsheet files to be used for the analysis phase. The analysis may be one of the following.

(a) Interannotation agreement.

(b) Graph analysis from detected entities.

(c) Cross-document analysis with other databases.

All these features are shown in the java GUI buttons, right click menu, and keyboard shortcuts that make SAMNA a user-friendly tool.

REFERENCES

Alawieh, A., Sabra, Z., Sabra, M., Zaraket, F., 2015a. Novel bioinformatics approach reveals pathogenic mechanisms in cerebral ischemia—a step towards preclinical stroke information management system. Stroke 46 (Suppl. 1), AWP408.

Alawieh, A., Sabra, Z., Sabra, M., Zaraket, F., 2015b. In: A rich club organization in spinal cord injury interactome provides insight into pathophysiological mechanisms and potential therapeutic interventions. Spinal Cord Injury, ASCI/AAP Joint Meeting, Chicago.

Björne, J., Salakoski, T., 2013. TEES 2.1: automated annotation scheme learning in the BioNLP 2013 shared task. Proceedings of the BioNLP Shared Task 2013 Workshop, pp. 16–25.

Day, D.S., McHenry, C., Kozierok, R., Riek, L.D., 2004. In: Callisto: a configurable annotation workbench. LREC.

Kolb, P., 2009. Experiments on the difference between semantic similarity and relatedness. Proceedings of the 17th Nordic Conference of Computational Linguistics (NODALIDA 2009), pp. 81–88.

Morton, T., Lacivita, J., 2003. WordFreak. Proceedings of the 2003 Conference of the North American Chapter of the Association for Computational Linguistics on Human Language Technology Demonstrations—NAACL 03. https://doi.org/10.3115/1073427.1073436.

Müller, C., Strube, M., 2006. Multi-level annotation of linguistic data with MMAX2. In: Corpus Technology and Language Pedagogy: New Resources, New Tools, New Methods. vol. 3, Peter-Lang-Verlagsgruppe, Bern, Switzerland, pp. 197–214.

Nédellec, C., Bossy, R., Kim, J.D., Kim, J.J., Ohta, T., Pyysalo, S., Zweigenbaum, P., 2013. In: Overview of BioNLP shared task 2013. Proceedings of the BioNLP Shared Task 2013 Workshop.

Ogren, P.V., 2006. Knowtator. Proceedings of the 2006 Conference of the North American Chapter of the Association for Computational Linguistics on Human Language Technology companion volume: demonstrations. https://doi.org/10.3115/1225785.1225791.

Stenetorp, P., Pyysalo, S., Topić, G., Ohta, T., Ananiadou, S., Tsujii, J.I., 2012. BRAT: A Web-Based Tool for NLP-Assisted Text Annotation. In: Proceedings of the Demonstrations at the 13th Conference of the European Chapter of the Association for Computational Linguistics. Association for Computational Linguistics, Chicago, pp. 102–107.

Szklarczyk, D., Franceschini, A., Kuhn, M., Simonovic, M., Roth, A., Minguez, P., Doerks, T., Stark, M., Muller, J., Bork, P., Jensen, L.J., Mering, C.V., 2010. The STRING database in 2011: functional interaction networks of proteins, globally integrated and scored. Nucleic Acids Res. https://doi.org/10.1093/nar/gkq973.

The UniProt Consortium, 2010. The Universal Protein Resource (UniProt) in 2010. Nucleic Acids Res. https://doi.org/10.1093/nar/gkp846.

Zaraket, F.A., Jaber, A., 2013. MATAr: morphology-based tagger for Arabic. 2013 ACS International Conference on Computer Systems and Applications (AICCSA). https://doi.org/10.1109/aiccsa.2013.6616418.

CHAPTER 11

Intractome Curation and Analysis for Stroke and Spinal Cord Injury Using Semiautomatic Annotations

Mohamed Sabra*, Ali Alawieh†, Fadi A. Zaraket*

*Department of Electrical and Computer Engineering, Maroun Semaan faculty of engineering and architecture, American University of Beirut, Beirut, Lebanon
†Department of Microbiology and Immunology, Medical University of South Carolina, Charleston, SC, United States

This chapter presents a methodology to use semiautomatic annotations and existing systems biology resources to construct intractome graphs concerning a given disease.

The methodology has the following processes.

1. Gene and protein accession and mapping process.
2. Article selection process.
3. Coocurrence relation extraction process.
4. Systems biology relations integration process.
5. Analysis process.

In the following, we describe each of the processes and discuss its application to the stroke and spinal cord injury diseases. Then we report on the results and the interpretation of the analysis for both diseases.

The work in this chapter was reported in Alawieh et al. (2015a,b) open access journals and is covered by the Creative Commons Attribution license (CC BY: https:// creativecommons.org/licenses/by/4.0/legalcode).

1. GENE AND PROTEIN ACCESSION AND MAPPING PROCESS

Gene and protein expressions have variant expressions and mentions referring to the same gene and protein entity. We used two resources for the accession and mapping of the gene and protein mentions. The first is the Universal Protein Resource (UniProt) and the second is the HUGO Gene Nomenclature Committee (HGNC) database. We retrieved human orthologues and corresponding accessions for each protein mention from both resources.

Leveraging Biomedical and Healthcare Data
https://doi.org/10.1016/B978-0-12-809556-0.00011-3

2. ARTICLE SELECTION PROCESS

The article selection process proceeds as follows:

1. The publication collection process starts with a specialist designating a mesh search across publication databases such as PubMed. For example, for the stroke disease, we used the "(Stroke OR brain infarct* OR cerebral infarct* OR brain ischem* OR cerebral ischem* OR ischemic brain injury)" mesh search query and retrieved 82,181 articles that discuss the stroke disease.

2. The article selection process filters the retrieved articles based on their inclusion of protein mentions. This is a semiautomatic process where we use automatic annotation to find protein and gene change expression mentions in the articles. Then we validate the annotations and proceed in several iterations until the resulting automatic annotation is satisfactory.

3. The semiautomatic process produces three sets of articles: (1) excluded articles, (2) included articles, and (3) undecided articles. The third group is typically a small minority that we manually inspect and make a decision to include some of its relevant articles in constructing the intractome.

4. We feed the included group of articles to a learning process. The learning process may automatically select a set of the rest of the articles for further consideration.

The flow diagram in Fig. 1 shows the application of the article selection process to the stroke disease. A total of 82,181 articles matched the mesh search on PubMed, 72,028 of

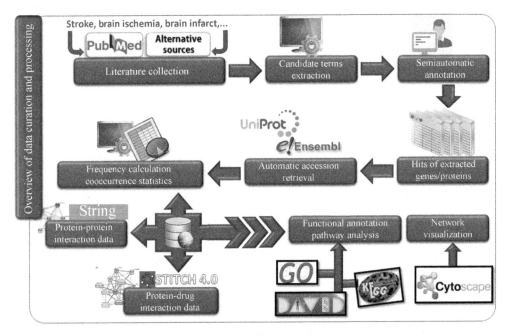

Fig. 1 Article selection process using semiautomatic annotations.

them did not contain protein and gene expression changes, 8948 articles contained clear expression changes, and 1205 articles were not decisive. We performed full text screening for the latter articles and included 115 of them to have a total of 9063 articles selected. The automatic annotation process results in 10,120 abstracts with annotated terms; 8043 of them were common with the selected articles, and 2377 were not. We performed manual screening for the abstracts and full texts of those latter articles and selected an additional 337 articles. This resulted in a total of 8740 articles selected.

Fig. 2 illustrates the gene and protein accession and mapping process applied to the selected 8740 articles. We applied the accession mapping extracted from UniProt and HGNC in addition to expert rules that represent possible term variability across abstracts. Based on the validated annotations of the 8740 abstracts, we concluded that 3% of the annotated terms had ambiguous annotations, 21% of the annotated terms had valid multiple corresponding accessions, and 76% of the annotated terms mapped to a single corresponding accession. We further inspected the 24% of the annotated terms to resolve the ambiguity as needed. An automatic process eliminated possible accessions with low probabilities, and then manual screening resolved the rest of the ambiguities.

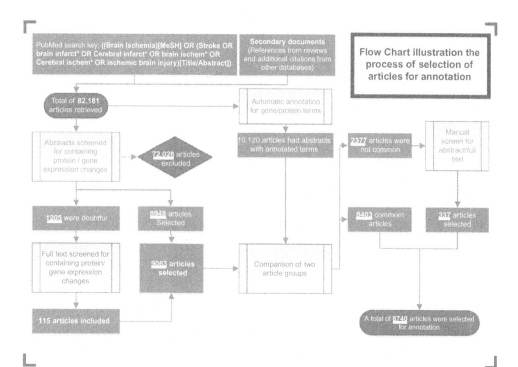

Fig. 2 Gene accession and mapping process.

3. COOCCURRENCE RELATION EXTRACTION PROCESS

The information extraction phase of our methodology starts with extracting the annotations from the text as atomic entities. Each annotation entity is linked to its context with a neighborhood relation to other entities. Annotations with the same label are characterized with statistical features such as their frequency and their colocation distribution with respect to other entities and other words and phrases in the text. This information is used as follows:

1. We use distributional similarity to suggest words and phrases that were not annotated in the first phase as entities.
2. We hypothesize that a rhetoric semantic relation exists between entities that occur in a given neighborhood. For simplicity and without loss of generalization, we consider binary colocation relations and form a relational entity whenever the hypothesis stands.
3. We characterize the relational entities with their distribution with respect to other entities and words and phrases in the text that were not annotated.
4. We use distributional similarity to find nonannotated words and phrases that have similar distributions to the relational entities. We suggest these words and phrases as relational evidence and use them to form relational entities iteratively.
5. We repeat the process until no changes are introduced or until a user-specified bound is met.

The result of the cooccurrence is a set of entities E and a set of relational entities $RE \subseteq E \times E \times W$ where W is the set of phrases occurring between the entities. A relational entity $re = \langle e_1, e_2, w \rangle \in RE$ has a phrase $w \in W$ occurring between e_1 and e_2.

These entities and relational entities form the seed of our intractome graph $IG = (E, RE)$ where elements of E are the vertices of the graph and elements of RE are the edges of the graph.

4. SYSTEMS BIOLOGY RELATIONAL EXTRACTION PROCESS

Systems biology databases with protein-protein interaction relations as well as drug-gene and drug-protein interaction relations exist. We used the STRING online database that contains curated protein-protein interaction data from several other databases (Franceschini et al., 2013). We also used the GeneCodis database to extract enriched information about gene-drug interactions within a network based on the PharmGKB knowledge base (Huang et al., 2008). Additionally, we used chemical-gene interaction information from the STITCH database that contains a network of more than 3 million chemical agents (Kuhn et al., 2011). The interactions in STITICH and GeneCodis are relational entities where the drug of the chemical agent is a node and the target of the interaction is either a gene or a protein.

We also used the DAVID online database to extract enriched gene ontology (GO) biological processes, cellular components, and tissue expressions (Huang et al., 2008). We also used the KEGG collection of databases for pathway relations (Kuhn et al., 2011). These relations link proteins to each other in what we can use later for clustering and functional annotation.

Formally, we consider a relation RE_{DB} that relates entities from a database DB. We project that relation over the selected set of disease entities E to form a disease network from the database $RE_{DB}^E = \{re \cdot re \in RE_{DB} \text{ and } re \cdot e_1 \in E \text{ or } re \cdot e_2 \in E\}$. The final disease intractome graph $DIG = \cup_{db \in \{IG, STRING, GENECODIS, STITCH, DAVID, KEGG\}} RE_{db}^E$ is the union of all disease networks across all databases with the seed intractome IG.

The stroke disease with included relations from STRING with above 0.4 combined score resulted in a stroke intractome of 886 protein nodes with 17,425 relations. The enrichment of the intractome with relations from DAVID and GO showed a dominance of brain tissue expressions within the proteins. Fig. 3 shows that 441 proteins have a brain tissue type. The enrichment score of the inflammation and response to injury biological processes came also with the highest score of 60 followed with homeostatic mechanisms, response to estradiol, and regulation of score death biological processes, as depicted in Fig. 4. Functional annotation of the stroke intractome revealed that the majority of the proteins are present in the extracellular space and plasma membrane compared to

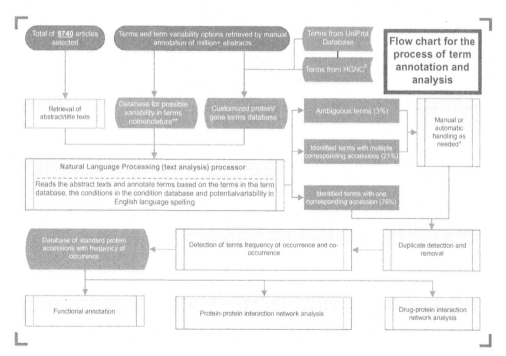

Fig. 3 Gene ontology annotation and tissue expression.

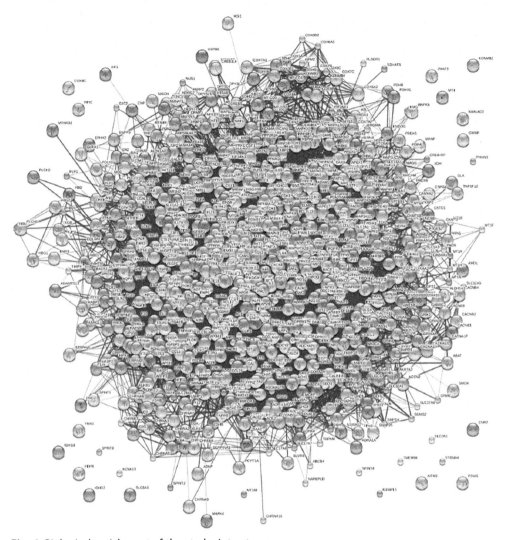

Fig. 4 Biological enrichment of the stroke intractome.

cytosol and cellular fractions, indicating that the majority of the pathophysiological events after stroke occur on and around the cell surface, as depicted by Fig. 5.

The pathway enrichment from KEGG considered pathways with a *P*-value less than 10^{-12}. Fig. 6 shows that the most enriched pathway was the complement and coagulation cascade (CCC). Together with the calcium signaling and MAPK signaling pathways, the three form the most significant pathways in the intractome. CCC has only a 4.4% overlap with other pathways compared to MAPK and calcium signaling, which featured 22% and 48% overlap, respectively.

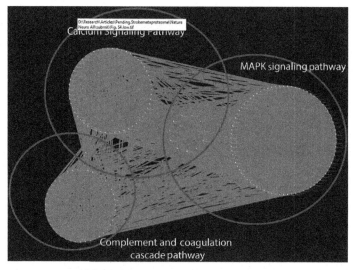

Fig. 5 Functional annotation of the stroke intractome.

5. THE ANALYSIS PROCESS

The analysis process concerns applying graph analysis techniques to infer meaningful information and insight from the constructed graph. Some of the techniques follow:

1. Reachability analysis techniques check whether a node in the graph is accessible from another. The reachability could also be more constrained to have the path annotated with the same enriched functional annotation, for example.

2. Graph similarity techniques check what parts of the network are similar in structure and connectivity to each other. These techniques include graph isometric computations and may lead to discovering protein interactions that behave in a similar manner.

3. Graph clustering techniques identify nodes that minimize connectivity and flow when clustered together. Iterative clustering can be applied to discover a hierarchy of interactions or to discover nodes that behave in a similar fashion with respect to other clusters of nodes.

4. Graph centrality techniques allow the discovery of nodes that are centric to the network.

5. Graph connectivity techniques allow the discovery of nodes that are essential to the connectivity of the network. These could be articulation nodes whose removal may directly isolate parts of the network. They could also be small subsets of nodes whose union is enough to connect the whole network.

For both stroke and spinal cord injury, we examined the topology of the extracted intractomes using the Systems Biology and Evolution MATLAB Toolbox (SBEToolbox) (Kuhn et al., 2011) and Cytoscape (Konganti et al., 2013). We compute characteristic

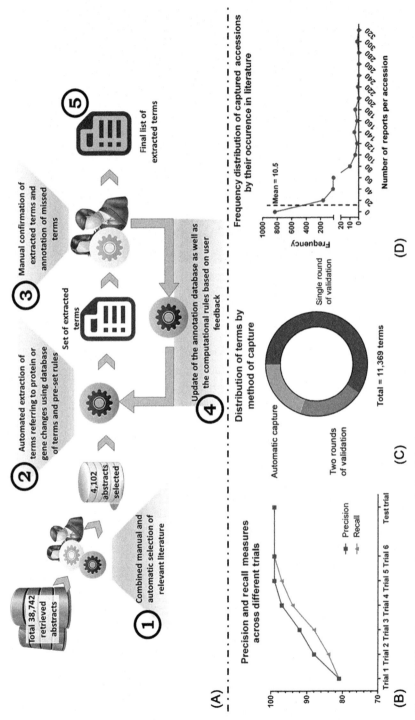

Fig. 6 KEGG pathway annotations and analysis of the intractome.

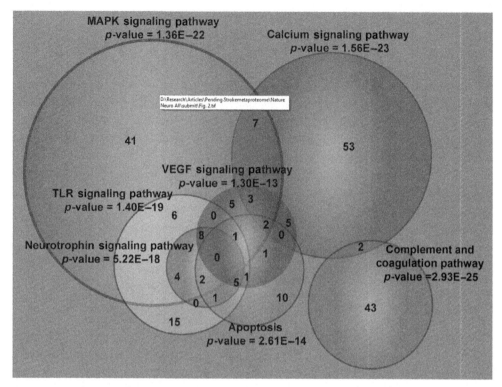

Fig. 7 CCC pathway in the stroke intractome.

measures of network organization node-specific degrees, clustering coefficients, path lengths, between-ness centrality, and modularity.

Fig. 7 shows how, for the stroke intractome, the CCC pathway is heavily interconnected with the other two prominent pathways, despite its low 4.4% overlap. This is significant given the early role of CCC in the recognition and response to ischemic and reperfusion injury (Elvington et al., 2012).

6. RICH-CLUB ANALYSIS

A rich club is a subset of network nodes with (1) high degrees, (2) highly interconnected among each other, and (3) more densely connected within the network than predicted by the node degrees alone (Colizza et al., 2006). A rich club is characterized by a rich-club coefficient $\phi(k)$ where k is a node degree. The coefficient is computed as the ratio of $E_{>k}$ the number of connections present between nodes with degree bigger than k to the total number of possible connections between those nodes. The latter is given by $N_{>k}(N_{>k}-1.)$

Where $N_{>k}$ is the number of nodes with degree higher than k.

$$\phi(k) = \frac{2E_{>k}}{N_{>k}(N_{>k}-1)}$$

The metric is a monotonically increasing metric even for randomly generated networks. So it cannot be directly used to imply the existence of a rich club organization in the intractome as in certain high degree distributions, it is not possible to avoid connecting high degree hub nodes. Consequently, we compute a normalized metric that is the ratio of the rich-club coefficient of the intractome $\phi(k)$ to that of a maximally randomized network $\phi_{rand}(k)$ with the same degree distribution $P(k)$. The presence of a rich-club organization is denoted by a ratio $\phi(k)/\phi_{rand}(k)$ that is bigger than 1 (Viger and Latapy, 2005).

Both intractomes of stroke and spinal cord injury showed the presence of a rich-club network. We review both diseases and report on their intractome analysis in the following sections.

7. STROKE CASE STUDY

The burden of ischemic stroke is still the highest among all neurological diseases, despite tremendous efforts devoted to prevention, management, treatment, and rehabilitation of stroke patients (Whiteford et al., 2016; Roger et al., 2012). After decades of research and hundreds of clinical trials on ischemic stroke, the full spectrum of pathophysiological processes has not yet been elucidated, nor has a final and terminal treatment been devised. Preclinical and clinical studies have predicted that a single-action, single-target paradigm is not the optimal approach to treat stroke and that multiaction, multitarget paradigms are required (Zlokovic and Griffin, 2011). Eventually, there is a sincere need to compile efforts to understand the evolution of different mechanisms after ischemic stroke and the relation of this to disease outcome and possible interventions.

8. STROKE INTRACTOME NETWORK ANALYSIS RESULTS

Fig. 8 shows that the frequency of nodes with certain degree (k) is inversely correlated with the degree (k), indicating that a few nodes have the majority of the interactions in the network and are thus hub nodes (illustrated in Fig. 8, right). This is consistent with scale-free networks, a property of most biological networks that exhibits a power-law degree distribution. This suggests a small world organization presence in the network that was verified by a higher clustering coefficient to random networks while having a comparable path length.

Fig. 9 shows the raw rich-club coefficient of the stroke intractome network in blue compared to that of the maximal random network in red. The normalized rich-club

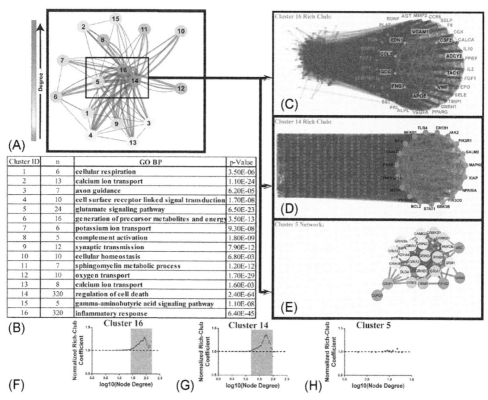

Cluster ID	n	GO BP	p-Value
1	6	cellular respiration	3.50E-06
2	13	calcium ion transport	1.10E-24
3	7	axon guidance	6.20E-05
4	10	cell surface receptor linked signal transduction	1.70E-08
5	24	glutamate signaling pathway	6.50E-23
6	16	generation of precursor metabolites and energy	3.50E-13
7	6	potassium ion transport	9.30E-08
8	5	complement activation	1.80E-09
9	12	synaptic transmission	7.90E-12
10	10	cellular homeostasis	6.80E-03
11	7	sphingomyelin metabolic process	1.20E-12
12	10	oxygen transport	1.70E-29
13	8	calcium ion transport	1.60E-03
14	320	regulation of cell death	2.40E-64
15	5	gamma-aminobutyric acid signaling pathway	1.10E-08
16	320	inflammatory response	6.40E-45

Fig. 8 Markov clustering of the network of BII, and the Power-law degree distribution for stroke intractome.

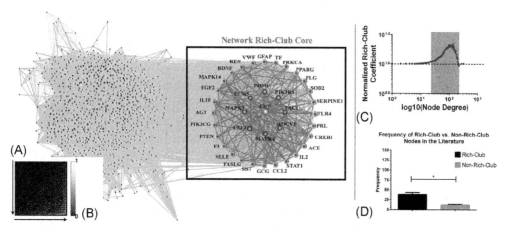

Fig. 9 Rich-club and normalized rich-club coefficients for the stroke intractome network compared to that of the maximal random network.

coefficient ρ is shown in green. The region between degrees 40 and 180 with a peak at degree 132 indicates the rich-club organization region of the network. The strongest rich-club component region is highlighted in red. Interestingly, only one node, C-reactive protein (CRP), had the degree 132.

Interestingly, a comparison of rich-club components to nonrich-club components for frequency of encounters in the curated literature shows that the frequency of rich-club nodes is significantly higher. Members of the rich club were found to span multiple pathophysiological pathways that predominantly include inflammatory and immunological response mechanisms.

The analysis of the enrichment of the intractome with drug-protein interaction from STITCH (Kuhn et al., 2011; Tabas-Madrid et al., 2012) revealed that estrogen is the most-enriched chemical therapeutic within the stroke intractome. Targets of estrogen include up to 15% of the nodes in the network. Estrogen was also found to preferentially target nodes within the rich club (53% of total rich-club components), which was reflected by a significant enrichment on the Fischer Exact t-test. Eventually, estrogen targets were shown to have a significantly higher normalized rich-club coefficient compared to estrogen nontargets. Besides estrogen, other chemical compounds that have a similar pleiotropic effect (beneficial or harmful) on targeting components of the intractome include nitric oxide and its donor L-arginine, ATP, tacrolimus, and glucocorticoids. More analysis and discussion of the stroke intractome are detailed in Alawieh et al. (2015a).

9. SPINAL CORD INJURY INTRACTOME CONSTRUCTION AND ANALYSIS

Spinal cord injury is a prominent cause of disability worldwide. It is second in the United States only to stroke as a cause of disability, accounting for 23% of all cases of paralysis (Gibson et al., 2009). The fact that there is no effective therapeutic intervention for the treatment of SCI highlights the need for a better and more integrative understanding of the molecular mechanisms that promote pathology and determine recovery.

There is limited information available on how pathways involved in these processes connect with each other to result in the pathological outcome. With the aim of providing insight into the interdependency of prominent pathophysiological processes and the molecular disturbances that affect neuronal survival and axonal regeneration, we investigated the full molecular architecture of SCI using our methodology.

We applied our methodology to SCI and started with a mesh term search query ("Traumatic Spinal Cord"[Title/Abstract] OR "Spinal Cord Trauma"[Title/Abstract]]) to extract articles from PubMed and from references of reviews and extracted papers. We collected 38,742 articles and selected 4102 articles as a basis for the construction of the intractome. The process involved six iterations until 99.5 precision and recall measures were achieved.

We then captured a total of 11,369 terms that included automatically captured and manually verified terms. Captured terms were mapped to 1083 unique accessions for proteins or genes associated with SCI pathogenesis. The distribution of those accessions by occurrence in the literature was right skewed with a mean of 10.5 reports per accession.

The SCI intractome exhibited both scale-free and small-world properties with a low path length of 2.45 and a high clustering coefficient of 0.42. The SCI intractome also contained a rich-club organization.

In comparison, components in the SCI rich club were twice more frequently studied in the literature than nonrich club components.

The most significantly enriched pathways in the rich club (P-value $<1E-10$) have 100% overlap with the top 10 pathways in the full network. This indicates that the rich club involves the major hubs of the key pathways in the network.

Further analysis showed dense interconnections between the different rich-club pathways centered on a densely connected core of pathways involved in the final death or survival decision in the cell. Peripheral pathways involve those responsible for recognition of microenvironmental changes, including injury detection (complement and coagulation cascade, NOD-like receptor signaling, Fc-epsilon receptor signaling), growth factor recognition (VEGF and neuroactive ligand signaling), and response to extracellular changes (calcium signaling, gap junctions). These peripheral pathways are less densely interconnected and serve as relays of microenvironment information to the central core of signal transduction pathways to favor an ultimate outcome of neuronal degradation or survival.

We assessed whether the SCI intractome exhibited a significant knotty center in addition to the rich-club organization. While rich-club centrality is based on node-specific degree measures, knotty centrality captures central subnetworks that are highly connected to the whole network while the nodes in the subnetwork themselves may not be of high degrees (Shanahan and Wildie, 2012).

We performed knotty centrality on both the entire network and the rich club of the SCI intractome. Fig. 10A shows the most significant knotty center subnetworks of the SCI intractome and Fig. 10B shows that of the SCI intractome rich club.

There is approximately 67% overlap across both the network knotty center and the rich-club knotty center. Therefore, we infer that the SCI intractome rich club is central to the connectivity of the entire network: it is a highway system for communication of nonrich-club nodes.

We validated this conclusion via collapsing the rich-club nodes into their modules. This dilutes the effects of the intrarich-club connectivity. Then we checked for the presence of a knotty center in the collapsed network of nonrich-club nodes and rich-club muddle interactions. The majority (65%) of the knotty center we found was composed of the rich-club modules. The analysis of individual nodes in the intractome and more discussion is detailed in Alawieh et al. (2015b).

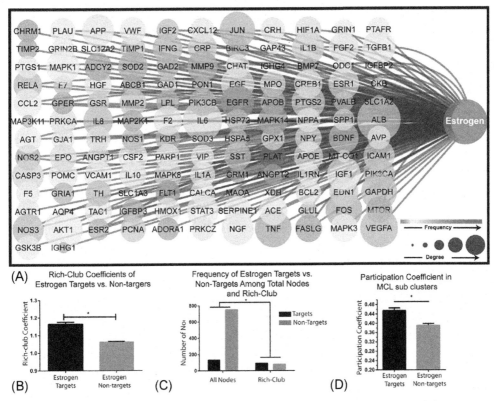

Fig. 10 Knotty centrality for the SCI intractome and the rich-club SCI intractome, and the estrogen targets within the BIMP.

10. CONCLUSION

We used SAMNA to construct, annotate, and curate an intractome for the stroke disease and an intractome for the SCI disease. We enriched both intractomes with systems biology gene ontology, functionality, and interaction databases. We inspected the resulting enriched intractomes with graph techniques and inferred important and significant findings that are not possible to reach with a human-based approach. For the stroke disease, the curation of the intractome allowed for understanding the role of biochemical pathways in the amplification of injury after stroke. Graph techniques revealed the presence of a rich-club organization. The detected pathways and the rich club provided insight on how to target injury mechanisms after stroke. We characterized the full molecular architecture of SCI. Graph techniques revealed a modular organization of the disease process that is centered on a core of cell survival decision proteins. Those core proteins provide the best targets for therapy.

REFERENCES

Alawieh, A., Sabra, Z., Sabra, M., Tomlinson, S., Zaraket, F.A., 2015a. A rich-club organization in Brain ischemia protein interaction network. Sci. Rep 5, 13513

Alawieh, A., Sabra, M., Sabra, Z., Tomlinson, S., Zaraket, F.A., 2015b. Molecular architecture of spinal cord injury protein interaction network. PLoS One 10, e0135024

Colizza, V., Flammini, A., Serrano, M.A., Vespignani, A., 2006. Detecting rich-club ordering in complex networks. Nat. Phys. 2, 110–115.

Elvington, A., et al., 2012. Pathogenic natural antibodies propagate cerebral injury following ischemic stroke in mice. J. Immunol 188, 1460–1468

Franceschini, A., Szklarczyk, D., Frankild, S., Kuhn, M., Simonovic, M., Roth, A., Lin, J., Minguez, P., Bork, P., Mering, C.V., Jensen, L.J., 2013. STRING v9.1: protein-protein interaction networks, with increased coverage and integration. Nucleic Acids Res. https://doi.org/10.1093/nar/gks1094.

Gibson, C., Turner, S., Donnelly, M., 2009. One Degree of Separation: Paralysis and Spinal Cord Injury in the United States. Christopher and Dana Reeve Foundation, Short Hills, NJ.

Huang, D.W., Sherman, B.T., Lempicki, R.A., 2008. Systematic and integrative analysis of large gene lists using DAVID bioinformatics resources. Nat. Protoc. 4, 44–57.

Konganti, K., Wang, G., Yang, E., Cai, J.J., 2013. SBEToolbox: a Matlab toolbox for biological network analysis. Evol. Bioinforma. https://doi.org/10.4137/ebo.s12012.

Kuhn, M., Szklarczyk, D., Franceschini, A., Mering, C.V., Jensen, L.J., Bork, P., 2011. STITCH 3: zooming in on protein-chemical interactions. Nucleic Acids Res. https://doi.org/10.1093/nar/gkr1011.

Roger, V.L., Go, A.S., Lloyd-Jones, D.M., Benjamin, E.J., Berry, J.D., Borden, W.B., Bravata, D.M., Dai, S., Ford, E.S., Fox, C.S., Fullerton, H.J., Gillespie, C., Hailpern, S.M., Heit, J.A., Howard, V.J., Kissela, B.M., Kittner, S.J., Lackland, D.T., Lichtman, J.H., Lisabeth, L.D., Makuc, D.M., Marcus, G.M., Marelli, A., Matchar, D.B., Moy, C.S., Mozaffarian, D., Mussolino, M.E., Nichol, G., Paynter, N.P., Soliman, E.Z., Sorlie, P.D., Sotoodehnia, N., Turan, T.N., Virani, S.S., Wong, N.D., Woo, D., Turner, M.B., 2012. Executive summary: heart disease and stroke statistics— 2013 update: a report from the American Heart Association. Circulation. https://doi.org/10.1161/cir.0b013e3182456d46.

Shanahan, M., Wildie, M., 2012. Knotty-centrality: finding the connective core of a complex network. PLoS One 7, e36579

Tabas-Madrid, D., Nogales-Cadenas, R., Pascual-Montano, A., 2012. GeneCodis3: a non-redundant and modular enrichment analysis tool for functional genomics. Nucleic Acids Res. https://doi.org/10.1093/nar/gks402.

Viger, F., Latapy, M., 2005. Efficient and simple generation of random simple connected graphs with prescribed degree sequence. In: Lecture Notes in Computer Science Computing and Combinatorics. https://doi.org/10.1007/11533719_45.

Whiteford, H.A., Ferrari, A.J., Degenhardt, L., Feigin, V., Vos, T., 2016. Global burden of mental, neurological, and substance use disorders: an analysis from the global burden of disease study 2010. In: Disease Control Priorities, Third Edition (Volume 4): Mental, Neurological, and Substance Use Disorders. https://doi.org/10.1596/978-1-4648-0426-7_ch2.

Zlokovic, B.V., Griffin, J.H., 2011. Cytoprotective protein C pathways and implications for stroke and neurological disorders. Trends Neurosci. 34, 198–209.

FURTHER READING

Demchak, B., Hull, T., Reich, M., Liefeld, T., Smoot, M., Ideker, T., Mesirov, J.P., 2014. Cytoscape: the network visualization tool for GenomeSpace workflows. F1000Res. https://doi.org/10.12688/f1000research.4492.1.

Huang, D.W., Sherman, B.T., Lempicki, R.A., 2009. Bioinformatics enrichment tools: paths toward the comprehensive functional analysis of large gene lists. Nucleic Acids Res. 37, 1–13.

Mering, C.V., 2003. STRING: a database of predicted functional associations between proteins. Nucleic Acids Res. 31, 258–261.

Shannon, P., Markiel, A., Ozier, O., Baliga, N.S., Wang, J.T., Ramage, D., Amin, N., Schwikowski, B., Ideker, T., 2003. Cytoscape: a software environment for integrated models of biomolecular interaction networks. Genome Res. 13, 2498–2504.

Sherman, B.T., Huang, D., Tan, Q., Guo, Y., Bour, S., Liu, D., Stephens, R., Baseler, M.W., Lane, H.C., Lempicki, R.A., 2007. DAVID knowledgebase: a gene-centered database integrating heterogeneous gene annotation resources to facilitate high-throughput gene functional analysis. BMC Bioinformatics 8, 42.

Szklarczyk, D., Franceschini, A., Kuhn, M., Simonovic, M., Roth, A., Minguez, P., Doerks, T., Stark, M., Muller, J., Bork, P., Jensen, L.J., Mering, C.V., 2011. The STRING database in 2011: functional interaction networks of proteins, globally integrated and scored. Nucleic Acids Res. https://doi.org/10.1093/nar/gkq973.

The UniProt Consortium, 2010. The Universal Protein Resource (UniProt) in 2010. Nucl. Acids Res. https://doi.org/10.1093/nar/gkp846.

CHAPTER 12

Deep Genomics and Proteomics: Language Model-Based Embedding of Biological Sequences and Their Applications in Bioinformatics

Ehsaneddin Asgari*, Mohammad R.K. Mofrad*,†
*Molecular Cell Biomechanics Laboratory, Department of Bioengineering and Department of Mechanical Engineering, University of California, Berkeley, Berkeley, CA, United States
†Physical Biosciences Division, Lawrence Berkeley National Lab, Berkeley, CA, United States

1. INTRODUCTION

What is the language of life?

Noam Chomsky introduces a broad definition for language, describing it as "a (finite or infinite) set of sentences, each finite in length, and constructed out of a finite set of elements" (Chomsky, 1957). Obviously, this definition is not limited to natural languages. An abstract representation of the internal dispositions of the macromolecules of life, that is, nucleic acids (DNA/RNA) and proteins, satisfies this definition as well, as all of these macromolecules are polymers constructed from a finite set of smaller molecules (Kuriyan et al., 2012). DNA and RNA are polymers made up of sequences of nucleotides of four distinct types with alphabetic representations {A, T, C, G} and {A, U, C, G} respectively, and proteins are polymers made up of sequences of amino acids of 20 different types represented by alphabet characters {A, C, D, E, F, G, H, I, K, L, M, N, P, Q, R, S, T, V, W, Y}. DNA and RNA are informational molecules as they carry the genetic instructions needed to make proteins. In contrast to informational molecules, proteins are operational macromolecules as they contribute to the molecular machinery that carries out the functions that are essential for life. The central dogma of molecular biology describes the relation among DNA, RNA, and proteins along the flow of genetic information. It states that information flows from the DNA into the RNA in a process called transcription, and then further to proteins through a process called translation (Kuriyan et al., 2012). Even in the terminology, the presence of elements such as transcription and translation reflects the notion of viewing this information representation system as the "language" of life.

What is the relationship between natural languages and the language of life?

Leveraging Biomedical and Healthcare Data
https://doi.org/10.1016/B978-0-12-809556-0.00012-5

Linguists and computational linguists consider a sentence as the output of a complex generative process controlled by certain rules. They distinguish between syntactic and semantic rules. Generally speaking, syntactic rules govern how the elements are put together to generate well-formed sentences while semantic rules determine the meaning of the resulting sentence. Analogous to what linguists and computational linguists believe about the sequence of words in a sentence, biologists believe that protein and nucleotide (DNA/RNA) sequences are not merely one-dimensional strings of symbols. These sequences encode a lot of information about molecular structure and functions in themselves (Kuriyan et al., 2012). Structures and functions of macromolecules are interesting for us as they can provide information about genotypes, phenotypes, diseases, or even treatments of diseases. Similar to the complex syntax and semantics of natural languages, certain biophysical and biochemical grammars dictate the formation of biological sequences. Thus, it would be natural to adopt/develop methods in natural language processing (also known as computational linguistics) to gain a deeper understanding of how functions and information are encoded within biological sequences (Yandell and Majoros, 2002; Searls, 2002; Asgari and Mofrad, 2015).

What is life language processing?

In this article, we propose computational linguistic modeling of biological sequences. This task is distinguished from approaches that have suggested the modeling of sequences using formal language theory (Dong and Searls, 1994; Muggleton et al., 2001; Searls, 2013). In this work, we focus on recent advances in deep learning for language processing (Collobert et al., 2011; Mikolov et al., 2013a). Although thanks to the progresses in genomic sequencing, a large number of biological sequences are available for analysis, due to the costs of crystallography and experiments, the number of known structure/functions (metadata) is not comparable with the large number of known biological sequences (raw data). Therefore, it would be extremely beneficial to infer information about structure/functions solely based on the sequence data. In this regard we have considered the following aims:

Computational linguistics representation learning for biological sequences

Our first goal is to implicitly extract information about the syntax and semantics of biological sequences using the large amount of available sequence data (raw data) and use it in downstream tasks, where there is a lack of sequences with metadata (Asgari and Mofrad, 2015). In this regard, we propose learning a general-purpose data representation format from the raw data for the downstream bioinformatic tasks. The significance of this task is detailed in the next section.

Computational linguistics comparison of genomic language variations

Our second goal is to quantify the distances between syntactic and semantic features of two genomic language variations (Asgari and Mofrad, 2016). This genomic variation can be related to linguistic differences in genomic sequences of different species or genomic sequences of different health conditions. We propose the use of the method suggested above for performing such a comparison.

Computational linguistics representation learning for biological sequences

We explained the relationship between natural languages and biological sequences as well as our approach on adopting language processing methods for biological sequences. Thus, to motivate the role of representation learning for biological sequences, we need to comment on its significance for computational processing of natural languages.

Can a computer automatically understand a piece of English text to find documents with similar content or automatically translate a given document to French? These types of tasks constitute the area with which natural language processing (also known as computational linguistics) is mainly concerned. The purpose of natural language processing (NLP) is to design algorithms allowing computers to understand natural languages for performing specific tasks (e.g., information retrieval, machine translation, semantic analysis, etc.). When we want to discuss a complex concept with an audience unfamiliar with the topic, we model or represent the concept within a framework that is understandable for the audience. The same logic applies in presenting a natural language text to a computer. Computers are experts in dealing with numerical values, vectors, and matrices. Thus, the first step in NLP is to vectorize natural language text for computers.

Words are the input units of almost all NLP tasks. Therefore, to utilize machines for language processing, we need to find proper vector representations of words that are interpretable by machines. We expect such representations to preserve some indications of similarity and dissimilarity between words. For instance, when we search a phrase in a search engine we expect the machine to consider words "formula" and "equation" to be similar and consider them dissimilar to an irrelevant word such as "cuisine." Thus, we should attribute similar vector representations to the words formula and equation, dissimilar to the vector representation of cuisine. As a reminder, vector similarity can be calculated using operations in linear algebra (e.g., dot product, Euclidian distance, and cosine similarity). Of course semantic similarity is not the only consideration we have in NLP tasks. As an example, part-of-speech tagging is one of the routine NLP tasks where the goal is to label words with their syntactic part-of-speech (e.g., verb, adverb, etc.). Presumably when we want to perform part-of-speech tagging, we desire a vector representation incorporating syntactic similarities.

The performance of NLP tasks largely depends on the quality of data representation (also known as feature extraction/engineering) (Collobert et al., 2011; Bengio et al., 2013). Before the recent advances in deep learning, computational linguists used to manually incorporate their prior linguistic knowledge about the downstream task in their vectorized data representations to achieve better accuracy in NLP tasks (Collobert et al., 2011). Although such methods of manual representation engineering work for many applications, they have a number of drawbacks. First of all, it would be good to have a general framework for data representation for all NLP tasks and not to design task-specific features. Second, designing relevant features requires high degrees of domain knowledge while we want to minimize human intervention in artificial intelligence. In addition, such manually designed representations can be either incomplete or overcomplete. The advent of deep neural network algorithms allowed automatic encoding of data into a proper representation and introduced representation learning as a new field in itself in the realm of machine learning (Bengio et al., 2013). Recent works in the area of representation learning have proposed successful representations of data in computer vision, speech recognition, and natural language processing (Mikolov et al., 2013a; LeCun et al., 2015; Graves et al., 2013).

The large amount of known biological sequence data as well as the underlying biophysical syntactic and semantic structures in these sequences motivate learning data representation from the raw data. Such a representation can be used as a general-purpose representation and subsequently be utilized in any task of interest in bioinformatics.

In Section 2, we will explore the above-mentioned aims in detail and in Section 3, we will summarize the proposal to give an overview of the contributions and the future steps of the proposed research.

Computational linguistics comparison of genomic language variations

Classification of language varieties is one of the prominent problems in linguistics (Smith, 2016). The term language variety can refer to different styles, dialects, or even a distinct language. It has been a longstanding argument that strictly quantitative methods can be applied to determine the degree of similarity or dissimilarity between languages (Kroeber and Chrétien, 1937; Sankaran and Taskar, 1950; McMahon and McMahon, 2003). The methods proposed in the 1990s and early 2000s mostly relied on the utilization of intensive linguistic resources. For instance, similarity between two languages was defined based on the number of common cognates or phonological patterns according to a manually extracted list (Kroeber and Chrétien, 1937; McMahon and McMahon, 2003). Such an approach, of course, is not easily extendable to problems involving new languages. Recently, statistical methods have been proposed to automatically detect cognates (Berg-Kirkpatrick and Klein, 2010; Hall and Klein, 2010; Bouchard-Côté et al., 2013; Ciobanu and Dinu, 2014) and subsequently compare languages based on the number of common cognates (Ciobanu and Dinu, 2014).

The purpose of this aim is to define a quantitative measure of distance between languages in the broad definition for language, which includes languages of life. Such a metric should reasonably take both syntactic and semantic variability of languages into account. A measure of distance between languages can have various applications, including quantitative genetic/typological language classification, styles and genres identification, and translation evaluation. More importantly for life language processing, comparing the biological languages generating the genome in different genomic language variations can help in the classification and characterization of sequences of interests in terms of different genotypes, phenotypes, and diseases. This can potentially shed light on important biological facts as well.

2. MATERIAL AND METHODS

2.1 Training Distributed Representation for Sequence Segments

Continuous vector representations known as word vectors have recently become popular in natural language processing (NLP) as an efficient approach to represent semantic/syntactic units (Collobert et al., 2011; Mikolov et al., 2013a). Word vectors are trained in the course of training neural network-based language models from large amounts of textual data (words and their contexts) (Mikolov et al., 2013b). To be more precise, word representations are the outputs of the last hidden layers in neural networks trained for the prediction of the context of a given word, which is analogous to language modeling. Thus, word vectors are supposed to encode the most relevant features to language modeling by observing various samples. In such a representation, similar words have closer vectors, where similarity is defined in terms of both syntax and semantics. By training word vectors over large corpora of natural languages, interesting patterns have been observed. Words with similar vector representations display various types of similarity. For instance, **King-Man** + **Woman** is the closest vector to that of the word **Queen** (an instance of semantic regularities) and **quick − quickly ∼ slow − slowly** (an instance of syntactic regularities). Word vectors are used as general-purpose data representation methods in many natural language processing applications, including part-of-speech tagging, machine translation, and information retrieval (Collobert et al., 2011; Le and Mikolov, 2014; Li et al., 2015; Guo et al., 2014). In this project, we suggest adopting such representation for biological sequences.

We propose using word vector representations for segments (*n*-grams) of biological sequences, called biovectors (Asgari and Mofrad, 2015). Biovectors can be utilized as the representation method for a wide array of tasks in bioinformatics. In this work, for evaluation purposes we employed these features in protein family classification, intron-exon prediction, and domain identification, where high accuracies were obtained (Asgari and Mofrad, 2015).

For training word vectors in NLP, a large corpus should be used to ensure that a sufficient number of contexts have been observed for all words. Thus, in particular for protein sequences, we use Swiss-Prot as a rich protein database, which consists of 546,790 manually annotated and reviewed sequences (Boutet et al., 2016). A common feature extraction method in bioinformatics is splitting sequences in *n*-grams (Tomović et al., 2006; Osmanbeyoglu and Ganapathiraju, 2011; Mantegna et al., 1995); similarly here we split sequences into nonoverlapping *n*-grams, but we consider all possible ways of splitting for each sequence, as detailed in Asgari and Mofrad (2015).

2.1.1 Skip-Gram Neural Network

In training word vector representations, the skip-gram neural network attempts to maximize the average probability of contexts for given words in the training data:

$$\arg\max_{v,v'} \frac{1}{N} \sum_{i=1}^{N} \sum_{-c \leq j \leq c} \log p(w_{i+j} \vee w_i) p(w_{i+j} | w_i) = \frac{\exp\left(v'^{T}_{w_{i+j}} v_{w_i}\right)}{\sum_{k=1}^{W} \exp\left(v'^{T}_{w_{i+j}} v_{w_i}\right)},$$

where N is the length of the training, $2c$ is the window size we consider as the context, w_i is the center of the window, W is the number of words in the dictionary and v_w and v_w' are the n-dimensional word representation and context representation of word w, respectively. At the end of the training, the average of v_w and v_w' will be considered as the word vector for w. The probability $p(w_{i+j} | w_i)$ is defined using a softmax function. *However, the softmax calculation using all the vocabulary is computationally expensive. Thus, for the sake of efficiency, negative sampling technique has been used in Word2Vec implementation (Mikolov et al., 2013a; Goldberg and Levy, 2014).*

2.2 Evaluation of the Trained Distributed Representation

In natural language processing, researchers usually distinguish two categories of evaluations for word vectors: intrinsic evaluations versus extrinsic evaluations. An intrinsic evaluation examines the quality of representations (e.g., comparison of vector similarity with human judgment about similarity of words) independent of a specific NLP task while an extrinsic evaluation tests the strength of the representation in the downstream NLP tasks (Schnabel et al., 2015). Here we need to define intrinsic and extrinsic evaluations for our proposed representation.

2.2.1 Intrinsic Evaluation

Here we propose a new intrinsic evaluation for biovectors based on continuity of the underlying biophysical and biochemical properties in this space. To qualitatively analyze the distribution of various biophysical and biochemical properties within the training space, we project all 3-gram embeddings from 100-dimensional space to a 2D space using stochastic neighbor embedding (Maaten and Hinton, 2008). Mass, volume, polarity, hydrophobicity, charge, and van der Waals volume properties were analyzed. In addition, to quantitatively measure the continuity of these properties in the protein-space, the best Lipschitz constant, that is, the smallest k satisfying is calculated:

$$\arg\min_{k} d_f\left(f_{prop}(w_1), f_{prop}(w_2)\right) \leq k \times d_w(w_1, w_2)$$

2.2.2 Extrinsic Evaluation

For the purpose of extrinsic evaluation of biovector representation, we utilize the trained representation in a variety of bioinformatics tasks, including family classification, exon-intron prediction, domain identification, and protein-to-protein interaction prediction.

2.2.3 Protein Family Classification

A protein family is a set of proteins that is evolutionarily related, typically involving similar structures or functions. The large gap between the number of known sequences versus the amount of known functional information about sequences has motivated family (function) identification methods based on primary sequences (Bork et al., 1998; Enright et al., 2002). The Protein Family Database (Pfam) is a widely used source for protein families (Finn et al., 2015). In Pfam, a family can be classified as a "family," "domain," "repeat," or "motif."

The existing methods typically require extensive information for feature extraction, for example, hydrophobicity, normalized Van der Waals volume, polarity, polarizability, charge, surface tension, secondary structure, and solvent accessibility (Cai et al., 2003).

In this study, we utilize biovectors to classify protein families in Swiss-Prot using the information provided by the Pfam database, and we obtain a high classification accuracy (Asgari and Mofrad, 2015).

Each sequence is represented as the summation of the vector representation of overlapping 3-grams. Thus, each sequence is presented as a vector of size d ($d = 100$) (Asgari and Mofrad, 2015). Support vector machine classifiers are used to evaluate such a representation in the classification of protein families through a $10 \times$ fold cross-validation framework (Cortes and Vapnik, 1995).

2.2.4 Sequence Labeling

Many applications of statistical learning in natural language processing and bioinformatics can be considered as sequence labeling (e.g., part of speech tagging, semantic labeling, protein structure prediction, gene finding, etc.). Here, we want to evaluate the performance of biovector features in bioinformatics sequence by keeping the same learning algorithm and changing the data representation. We use the max-margin Markov network (M3Net) (Taskar et al., 2003) as one of the successful statistical approaches for structure prediction in machine learning. The reason for this is to benefit from the strength of graphical models in modeling the dependencies as well as the ability of max-margin Markov networks in the incorporation of kernels to deal with high-dimensional features, where we can use biovector features. This method is evaluated over intron-exon prediction and protein domain identification.

2.2.5 Intron-Exon Prediction

The term genome refers to the sequence of nucleotides that contains the genetic information. Some sections of the genome are functional in the sense that they can be translated to proteins (exonic regions) while other sections cannot be translated to proteins (introns). The rapid decrease in the cost of DNA sequencing and the large amounts of available data discourage us from manually annotating different regions in the genome. In addition, finding all the patterns that define an intronic or exonic region is not trivial.

Thus, this task can widely benefit from data-driven methods. A variety of methods have been proposed, including intuitive methods involving automation based on manually extracted patterns, hidden Markov models, support vector machines, and conditional random fields (Krogh, 1997; Bernal et al., 2007; Sonnenburg et al., 2007).

We perform intron-exon prediction using max-margin Markov networks to evaluate the performance of biovector data representation versus one-hot vector representation of amino acids. We use a dataset of 3000 DNA sequences with marked intron and exon regions from the exon-intron database (Shepelev and Fedorov, 2006). We use 80% of the sequences for training and 20% for evaluation.

2.2.6 Protein Domain Identification

Domains in proteins refer to regions in the primary sequences that can form a specific tertiary structure. Finding such regions is also an instance of sequence labeling. Protein domain identification is important, as they act as fundamental units of structure and function in a protein (Kuriyan et al., 2012; Murzin et al., 1995). Several methods have been proposed for predication of protein domains based on tertiary structure, multiple sequence alignment, and sequence-based approaches. Such approaches incorporate manually designed features including secondary structure, solvent accessibility, evolutionary profile, and amino acid entropy (Ingolfsson and Yona, 2008).

We use a dataset of 2340 protein sequences with the tyrosine kinase domain extracted from the SCOP database (Murzin et al., 1995). We use 80% of the sequences for training and 20% for evaluating the use of biovector data representation versus one-hot vectors in M3Net.

2.3 Word Embedding Language Divergence

We introduce a new quantitative measure of distance between genetic language variations based on word embedding, called word embedding language divergence, which is defined as the divergence between graphs of n-gram word vectors.

Our goal is to be able to provide a quantitative estimate of distance for any two given language variations, L and L'. In our framework, we define a language as a weighted graph $\Omega_L(V, e)$, where V is a set of vertices (n-grams), and $e: (V \times V) \rightarrow [-1, 1]$ is a weight function mapping a pair of words to their similarity value. Then our goal of approximating the distance between the two languages L and L' can be transferred to the approximation of the distance between $\Omega_L(V, e)$ and $\Omega_{L'}(V', e')$. We propose the use of cosine similarity between biovectors as the metric for n-gram similarities (weight function), which has been shown to take into account both syntactic and semantic similarities (Asgari and Mofrad, 2015; Mikolov et al., 2013a).

Because the sets of n-grams are the same within L and L' for biological sequences, to compare Ω_L and $\Omega_{L'}$ we only need to compare e and e'. We propose two methods for comparing e and e': we calculate the Jensen–Shannon divergence (JSD) between unified and normalized e and e':

$$JSD(\hat{e}_L \| \hat{e}_{L'}) = \frac{1}{2} D_{KL}(\hat{e}_L \| \bar{e}) + \frac{1}{2} D_{KL}(\hat{e}_{L'} \| \bar{e})$$

$$D_{KL}(\hat{e}_L \| \hat{e}_{L'}) = \sum_{i,j} \hat{c}_L(w_i, w_j) \log \frac{\hat{c}_L(w_i, w_j)}{\hat{c}'_L(w_i, w_j)}$$

We perform language comparison for the coding regions in the genomes of 12 different organisms using this method (4 plants, 6 animals, and 2 human subjects). The introns and exons were collected from the intron-exon database (Shepelev and Fedorov, 2006).

We train the word vectors for each setting of n-grams ($n = 3,4,5,6$) and organisms separately, using a skip-gram neural network implementation (Mikolov et al., 2013b). We set the number of dimensions for the word vector d to 100, and the window size c to 40. In addition, we subsample the frequent words by the ratio 10:3. Subsequently, we calculate language divergence with several different n-gram settings ($n = 3,4,5,6$) and two different divergence functions. This method is for classification of natural languages and genomic variations, as detailed in Asgari and Mofrad (2016).

3. RESULTS

3.1 Training a Distributed Representation for Sequence Segments

3.1.1 Intrinsic Evaluation Results

Although the biovectors are trained based on only the primary sequences of proteins, this offers several interesting biochemical and biophysical implications. In order to study these features, we visualized the distribution of different criteria, including mass, volume, polarity, hydrophobicity, charge, and van der Waals volume in this space. To do so, for each 3-gram we conducted qualitative and quantitative analyses, as described below.

3.1.2 Qualitative Analysis

In order to visualize the distribution of the aforementioned properties, we projected all 3-gram embedding from 100-dimensional space to a 2D space using stochastic neighbor embedding (t-SNE) (Maaten and Hinton, 2008). In the diagrams presented in Fig. 1, each point represents a 3-gram and is colored according to its scale in each property. Interestingly, as can be seen in the figure, 3-grams with the same biophysical and biochemical properties were grouped together. This observation suggests that the proposed embedding not only encodes protein sequences in an efficient way that proved useful for machine learning tasks, but also reveals some important physical and chemical patterns in protein sequences. Based on these results, we have showed that such a representation is useful for visualization and characterization of a set of sequences of interest as well (Asgari and Mofrad, 2015).

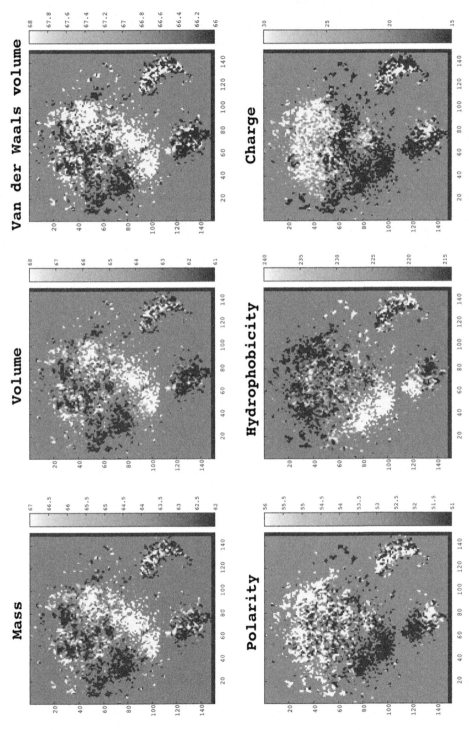

Fig. 1 Normalized distributions of biochemical and biophysical properties in biovector representation. In these plots, each point represents a 3-gram (a word of three residues) and the colors indicate the scale for each property. Data points in these plots are projected from a 100-dimensional space into a 2D space using t-SNE. As is shown, words with similar properties are automatically clustered together, meaning that the properties are smoothly distributed in this space.

Table 1 Using Lipschitz constant to evaluate the continuity of protein embeddings with respect to the biophysical and biochemical properties

Property	Lipschitz continuity		
	Embedding space	Scrambled space	Ratio
Mass	0.31	0.66	0.48
Volume	0.37	0.67	0.56
Van Der Waal volume	0.36	0.64	0.56
Polarity	0.48	1.26	0.38
Hydrophobicity	0.61	1.45	0.42
Charge	0.87	1.36	0.64
Average	0.50	1.01	0.51

3.1.3 Quantitative Analysis

Although Fig. 1 illustrates the smoothness of biovector representation with respect to different physical and chemical meanings, we required a quantitative approach to measure the continuity of these properties in this space. To do so, we calculated the best Lipschitz constant. For all six properties presented in Fig. 1, we calculated the minimum k. To evaluate this result, we made an artificial space called "scrambled space" by randomly shuffling the labels of 3-grams in the 100 dimensional space. Table 1 contains the values of Libschitz constants for biovector space versus the "scrambled space" with respect to different properties as well as their ratio.

3.2 Extrinsic Evaluation Results

3.2.1 Family Classification

In the classification of 324,018 protein sequences belonging to 7027 protein families using biovector features, an average family classification accuracy of $93\% \pm 0.06\%$ was obtained. The details and data can be downloaded from Harvard dataverse[1] (Asgari and Mofrad, 2015). A previous work reported accuracy in the range of 69.1%–99.6% for 54 protein families using manually designed features and a support vector machine classifier (Cai et al., 2003).

3.2.2 Sequence Labeling

Table 2 shows the results of the sequence labeling for intron-exon prediction and domain identification tasks. Our preliminary results show that incorporating the biovector features can increase the accuracy of sequence labeling in both tasks.

---

[1] https://doi.org/10.7910/DVN/JMFHTN.

Table 2 Comparison of intron-exon prediction and protein domain identification accuracies using Bio-vectors versus one-hot vector representations

Task	M3Net with one-hot vector features	M3Net with bio-vector features
Intron–exon prediction	73.84%	74.99%
Protein domain Identification	82.84%	89.8%

3.3 Word Embedding Language Divergence Results

The pairwise distance matrix for the 12 genetic languages with different n-gram settings ($n = 3,4,5,6$) is shown in Fig. 1. Our results confirm that evolutionarily closer species have a reasonably higher level of proximity in their language models. We can observe in Fig. 2 that as we increase the number of n-grams, the distinction between animal/human genomes and plants increases. This can be regarded as indicative of a high-level diversity between the genetic languages in plants versus animals. We have explored the classification of 50 natural languages using this method as well, which is detailed in Asgari and Mofrad (2016).

4. CONCLUSION

Finding a proper data representation that is interpretable by machines is an important step in any machine learning task. Distributed representations have proved to be a successful data representation approach in machine learning. We introduce a language model-based distributed representation method for biological sequences and train vector representations of sequence segments (n-grams), called biovectors, using skip-gram neural networks on large sequence databases. Biovectors can be utilized in feature extraction for a wide array of bioinformatics investigations such as family classification, structure prediction, disordered protein identification, gene finding, domain identification, and visualization. For evaluation purposes, we have employed these embeddings in the classification of 324,018 protein sequences belonging to 7027 protein families, where an average family classification accuracy of $93\% \pm 0.06\%$ was obtained. In addition, the incorporation of biovector features versus one-hot vector features in a Max-margin Markov Network for gene finding and domain identification tasks could improve the sequence labeling accuracy from 73.84% to 74.99% and from 82.4% to 89.8%, respectively. Biovectors are trained in the course of probabilistic language modeling. This makes the network of n-grams in this space a proper representation of the underlying language model. Considering this fact, we propose a new measure of distance between genomic language variations based on the divergence between embedding networks of n-grams in different genetic variations, called word embedding language divergence. We perform language comparison for the coding regions in the genomes of 15 different organisms (3 bacteria,

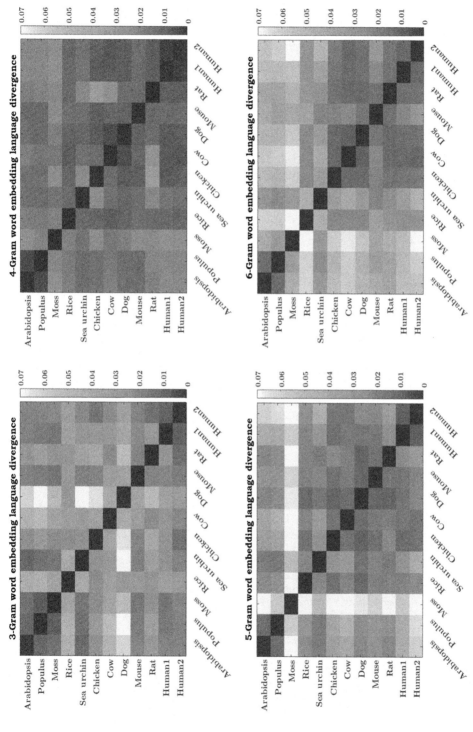

Fig. 2 Visualization of word embedding language divergence in 12 different genomes belonging to 12 organisms for various *n*-gram segments. As we increase the number of *n*-grams, the distinction between animal/human genomes and plants increases. Our results indicate that evolutionarily closer species have higher proximity in the syntax and semantics of their genomes.

4 plants, 6 animals, and 2 human subjects). Our results confirm a high-level difference in the genetic language model of humans/animals versus plants and bacteria. The proposed method is a step toward defining a new quantitative measure of similarity between genomic languages, with applications in characterization/classification of sequences of interest.

REFERENCES

Asgari, E., Mofrad, M.R.K., 2015. Continuous distributed representation of biological sequences for deep proteomics and genomics. PLoS One. 10. e0141287.

Asgari, E., Mofrad, M.R.K., 2016. In: Comparing fifty natural languages and twelve genetic languages using word embedding language divergence (WELD) as a quantitative measure of language distance. Multilingual and Cross-lingual Methods in NLP at NAACL-HLT. Association for Computational Linguistics, pp. 65–74.

Bengio, Y., Courville, A., Vincent, P., 2013. Representation learning: a review and new perspectives. IEEE Trans. Pattern Anal. Mach. Intell. 35, 1798–1828.

Berg-Kirkpatrick, T., Klein, D., 2010. In: Phylogenetic grammar induction.Proc. 48th Annu. Meeting (online).http://dl.acm.org/citation.cfm?id=1858812. [(Accessed 11 August 2016)].

Bernal, A., Crammer, K., Hatzigeorgiou, A., Pereira, F., 2007. Global discriminative learning for higher-accuracy computational gene prediction. PLoS Comput. Biol. 3, 0488–0497.

Bork, P., Dandekar, T., Diaz-Lazcoz, Y., Eisenhaber, F., Huynen, M., Yuan, Y., 1998. Predicting function: from genes to genomes and back. J. Mol. Biol. 283, 707–725.

Bouchard-Côté, A., Hall, D., Griffiths, T.L., Klein, D., 2013. Automated reconstruction of ancient languages using probabilistic models of sound change. Proc. Natl. Acad. Sci. U. S. A. 110, 4224–4229.

Boutet, E., Lieberherr, D., Tognolli, M., Schneider, M., Bansal, P., Bridge, A.J., Poux, S., Bougueleret, L., Xenarios, I., 2016. Uniprotkb/Swiss-prot, the manually annotated section of the uniprot knowledgebase: how to use the entry view. Methods Mol. Biol. 1374, 23–54.

Cai, C.Z., Han, L.Y., Ji, Z.L., Chen, X., Chen, Y.Z., 2003. SVM-Prot: web-based support vector machine software for functional classification of a protein from its primary sequence. Nucleic Acids Res. 31, 3692–3697.

Chomsky, N., 1957. Syntactic Structures. https://doi.org/10.1111/j.1467-9612.2004.00004.x.

Ciobanu, A., Dinu, L., 2014. In: An etymological approach to cross-language orthographic similarity. Application on Romanian.EMNLP. (online). http://www.aclweb.org/old_anthology/D/D14/D14-1112.pdf. [(Accessed 11 August 2016)].

Collobert, R., Weston, J., Bottou, L., Karlen, M., Kavukcuoglu, K., Kuksa, P., 2011. Natural language processing (almost) from scratch. J. Mach. Learn. Res. 12, 2493–2537.

Cortes, C., Vapnik, V., 1995. Support vector machine. Mach. Learn. https://doi.org/10.1007/978-0-387-73003-5_299.

Dong, S., Searls, D.B., 1994. Gene structure prediction by linguistic methods. Genomics 23, 540–551.

Enright, A.J., Van Dongen, S., Ouzounis, C.A., 2002. An efficient algorithm for large-scale detection of protein families. Nucleic Acids Res. 30, 1575–1584.

Finn, R.D., Coggill, P., Eberhardt, R.Y., Eddy, S.R., Mistry, J., Mitchell, A.L., Potter, S.C., Punta, M., Qureshi, M., Sangrador-Vegas, A., Salazar, G.A., Tate, J., Bateman, A., 2015. The Pfam protein families database: towards a more sustainable future. Nucleic Acids Res. https://doi.org/10.1093/nar/gkv1344.

Goldberg, Y., Levy, O., 2014. word2vec Explained: Deriving Mikolov et al.'s Negative-Sampling Word-Embedding Method. arXiv Prepr. arXiv1402.3722.

Graves, A., Mohamed, A., Hinton, G., 2013. Speech recognition with deep recurrent neural networks.IEEE Int. Conf. Acoust. Speech Signal Processing. https://doi.org/10.1109/ICASSP.2013.6638947.

Guo, J., Che, W., Wang, H., Liu, T., 2014. In: Revisiting embedding features for simple semi-supervised learning.Proc. 2014 Conf. Empir. Methods Nat. Lang. Processing.

Hall, D., Klein, D., 2010. Finding cognate groups using phylogenies.Proc. 48th Annu. Meeting (online). http://dl.acm.org/citation.cfm?id=1858786. [(Accessed 11 August 2016)].

Ingolfsson, H., Yona, G., 2008. Protein domain prediction. Methods Mol. Biol. 426, 117–143.

Kroeber, A., Chrétien, C., 1937. Quantitative classification of Indo-European languages. Language (Baltim). (online). http://www.jstor.org/stable/408715. [(Accessed 11 August 2016)].

Krogh, A., 1997. Two methods for improving performance of an HMM and their application for gene finding. Proc. Int. Conf. Intell. Syst. Mol. Biol. 5, 179–186.

Kuriyan, J., Konforti, B., Wemmer, D., 2012. The Molecules of Life: Physical and Chemical Principles. (online). https://books.google.com/books?hl=en&lr=&id=jwcPBAAAQBAJ&oi=fnd&pg=PP1&dq=the+molecules+of+life:+physical+and+chemical+principle&ots=NzT7a_DRSD&sig=HdPY9WdCFmAlXtu-TCCc0OPWRuk. [(Accessed 10 August 2016)].

Le, Q., Mikolov, T., 2014. In: Distributed representations of sentences and documents.Int. Conf. Mach. Learn.—ICML 2014vol. 32. pp. 1188–1196.

LeCun, Y., Bengio, Y., Hinton, G., 2015. Deep learning. Nature. (online). http://www.nature.com/nature/journal/v521/n7553/abs/nature14539.html. [(Accessed 13 November 2015)].

Li, C., Ji, L., Yan, J., 2015. In: Acronym disambiguation using word embedding.Proceedings of the Twenty-Ninth AAAI Conference on Artificial Intelligence, pp. 4178–4179.

Maaten, L.V.D., Hinton, G., 2008. Visualizing data using t-SNE. J. Mach. Learn. Res. 9, 2579–2605.

Mantegna, R.N., Buldyrev, S.V., Goldberger, A.L., Havlin, S., Peng, C.K., Simons, M., Stanley, H.E., 1995. Systematic analysis of coding and noncoding DNA sequences using methods of statistical linguistics. Phys. Rev. E 52, 2939–2950.

McMahon, A., McMahon, R., 2003. Finding families: quantitative methods in language classification. Trans. Philol. Soc.. (online). http://onlinelibrary.wiley.com/doi/10.1111/1467-968X.00108/full. [(Accessed 11 August 2016)].

Mikolov, T., Chen, K., Corrado, G., Dean, J., 2013a. In: Distributed representations of words and phrases and their compositionality.NIPS, pp. 1–9.

Mikolov, T., Corrado, G., Chen, K., Dean, J., 2013b. In: Efficient estimation of word representations in vector space.Proceedings of the International Conference on Learning Representations (ICLR 2013), pp. 1–12.

Muggleton, S.H., Bryant, C.H., Srinivasan, A., Whittaker, A., Topp, S., Rawlings, C., 2001. Are grammatical representations useful for learning from biological sequence data?—A case study. J. Comput. Biol. 8, 493–521.

Murzin, A.G., Brenner, S.E., Hubbard, T., Chothia, C., 1995. SCOP: a structural classification of proteins database for the investigation of sequences and structures. J. Mol. Biol. 247, 536–540.

Osmanbeyoglu, H.U., Ganapathiraju, M.K., 2011. n-Gram analysis of 970 microbial organisms reveals presence of biological language models. BMC Bioinformatics 12, 12.

Sankaran, C., Taskar, A., 1950. Quantitative classification of languages. Bull. Deccan. (online). http://www.jstor.org/stable/42929431. [(Accessed 11 August 2016)].

Schnabel, T., Labutov, I., Mimno, D., Joachims, T., 2015. In: Evaluation methods for unsupervised word embeddings.Proceedings of the 2015 Conference on Empirical Methods in Natural Language Processing, pp. 298–307.

Searls, D., 2002. The language of genes. Nature 420, 211–217.

Searls, D.B., 2013. Review: a primer in macromolecular linguistics. Biopolymers 99, 203–217.

Shepelev, V., Fedorov, A., 2006. Advances in the exon-intron database (EID). Brief. Bioinform. 7, 178–185.

Smith, A.D.M., 2016. Dynamic models of language evolution: the linguistic perspective. In: The Palgrave Handbook of Economics and Language, pp. 61–100. https://doi.org/10.1007/978-1-137-32505-1.

Sonnenburg, S., Schweikert, G., Philips, P., Behr, J., Rätsch, G., 2007. Accurate splice site prediction using support vector machines. BMC Bioinformatics 8, S7.

Taskar, B., Guestrin, C., Koller, D., 2003. Max-margin Markov networks. Adv. Neural Inf. Process. Syst. 16, 10.1.1.129.8439.

Tomović, A., Janičić, P., Kešelj, V., 2006. n-Gram-based classification and unsupervised hierarchical clustering of genome sequences. Comput. Methods Prog. Biomed. 81, 137–153.

Yandell, M.D., Majoros, W.H., 2002. Genomics and natural language processing. Nat. Rev. Genet. 3, 601–610.

CHAPTER 13

In Silico Transcription Factor Discovery via Bioinformatics Approach: Application on iPSC Reprogramming Resistant Genes

Natalia Polouliakh[*,†,‡]
[*]Sony Computer Science Laboratories Inc., Fundamental Research Laboratory, Tokyo, Japan
[†]Systems Biology Institute, Tokyo, Japan
[‡]Department of Ophthalmology and Visual Science, Yokohama City University Graduate School of Medicine, Yokohama, Japan

1. INTRODUCTION

Identifying the transcriptional signature is a powerful method for predicting cellular fate. However, combinatorial binding of transcription factors and long regulatory regions, which can include both the comparatively short promoter regions in the proximity of transcriptional start sites (TSS) and distantly located enhancer regions interacting with one or more promoter sites of neighboring genes (Serizawa et al., 2003), make unraveling the transcription regulation mechanism a challenging task. Additionally, transcription factor binding motif occurrences exhibit considerable variety, typically only partially matching the consensus pattern of the motif to which they belong (Polouliakh et al., 2004). Another difficulty is that because <5% of the human genome is encoding genes, the rest are covered with various repetitive elements, which are not transcription binding sites—and thus are a source of false positives for this task.

Microarray analysis provides us with new information about gene regulatory networks, and many novel candidates of coregulated genes are being identified. It is assumed that coregulated genes share *cis*-acting regulatory motifs in their upstream regions, and developing an algorithm for their accurate elucidation is still a big challenge. Discovering these regulatory elements can lead to the determination of functions and the establishment of evolutionary relationships among sequences. When the transcription factor binding motifs are unknown, the ab initio statistics methods are used. These methods can be divided into two major classes: methods based on word-counting (Sinha and Tompa, 2000; Hertz and Stormo, 1999) and methods based on probabilistic sequence models (Bailey and Elkan, 1995; Lawrence et al., 1993; Liu et al., 2001). The word methods analyze the frequency of oligonucleotides in the upstream region and use intelligent strategies to speed up counting and detect significantly overrepresented motifs.

Leveraging Biomedical and Healthcare Data
https://doi.org/10.1016/B978-0-12-809556-0.00013-7

These methods then compile a common motif by grouping similar words, which leads to a global optimum solution as it is implemented in the CONSENSUS (Stormo and Hartzell, 1989) and CoreSearch (Wolfertstetter et al., 1996) programs. The probabilistic methods represent the motif by a position-specific scoring matrix (PSSM), and the remainder of the sequence is modeled as a background model. Good examples of probabilistic programs are the Gibbs sampler (Lawrence et al., 1993), MEME (Bailey et al., 2015), MULTIPROFILER (Keich and Pevzner, 2002), and CONREAL (Berezikov et al., 2005). However, molecular biologists still encounter a serious problem when they have to decide which program should be chosen and how to maintain it for efficient operation (Polouliakh et al., 2004). One solution might be a "usage of all available" program because the relative superiority of each program varies according to the situation. For example, the MELINA (Poluliakh et al., 2003) program alleviates this problem somewhat by conveniently allowing *Users* to compare the results of four programs with various parameter settings.

Higher confidence in identification of cis-elements (motifs) in the regulatory regions of eukaryotic genomes can be achieved with supporting evidence such as cooccurrence of the same cis-elements in the existing orthologous promoters of other species. Among those methods, *phylogenetic footprinting* approaches such as Monkey (Moses et al., 2003) and Footprinter (Fang and Blanchette, 2006), and the *multiple alignment* approach, implemented in GPMiner (Lee et al., 2012) and SHOE (Polouliakh et al., 2004; Yoshino et al., 2016), can be distinguished.

The diversity of programs raises questions from *Users* as to which program is more suitable for a particular task. As running and installing tools are somehow problems for people without an informatics background, we investigated 31 online motif discovery tools useful for the comparative genomic analysis of biological pathways, in particularly the mTOR pathway (mammalian target of rapamycine) (Jablonska and Polouliakh, 2014), and identified that only nine were suitable for this kind of analysis. Other programs showed drawbacks in short promoter length, poor visualization, or web service failure. The diversity of existing approaches and data complexity let us assume that difficulties are not simply limited to the selection of the right motif discovery program, but to the selection of the right analytic path in order to reach the confident interpretation of the biological mechanism behind the dataset (Ghosh et al., 2011). Transcription regulation analysis on induced pluripotent stem cell (iPSC) reprogramming-resistant genes is described in this chapter.

2. REGULATION OF REPROGRAMMING RESISTANT GENES (RRG) BY iPSC REPROGRAMMING FACTORS Oct4, Sox2, KLF4, NANOG, AND c-Myc

The ultimate aim of research on induced pluripotent stem cells (iPSCs) is to create iPSCs that are identical to embryonic stem cells (ESCs) and that differentiate into tissue-specific cell types with intact functions (Fig. 1).

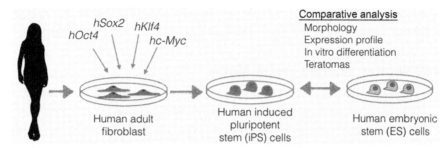

Fig. 1 Human dermal fibroblasts are exposed to viral vectors expressing a cocktail of four transgenes encoding four human transcription factors hOct4, hSox2, hKlf4, and hc-Myc.

Nevertheless, recognized discrepancies in gene expression between iPSCs and ESCs have been reported (Eckhardt et al., 2006; Chin et al., 2009; Muller et al., 2011). Comparison of expression signatures of 13 iPS and ES cell lines showed the existence of two groups of genes: iPSC reprogramming process-dependent genes, that is, "induced genes" (278 up- and 128 downregulated in more than two cell lines) and genes retained from somatic cells due to epigenetic memory "inherited genes" (1367 up- and 1113 downregulated in more than three cell lines) (Polouliakh, 2013; Gupta et al., 2010).

2.1 Somatic Cell Memory-Inherited Genes

Inherited genes can be considered as a part of transcriptional and epigenetic memory. They are of two origins: (a) those retaining their methylation status from somatic substrates, and (b) those activated or repressed through viral transduction in iPSCs in the course of reprogramming. Half the inherited genes are genes with univalent (H3K4, trimethylation of lysine residue 4) modification status in ES cell genes (P-value <0.001) when compared with the induced genes category by the chi-square independence test, pointing to their commitment to development. Reprogramming a somatic cell into a pluripotent state generates hundreds of aberrantly methylated loci at CpG islands, subsequently associated with genes (Meissner et al., 2008). Thus, this category of inherited genes, including many demethylated loci, is the most prone to reprogramming (Polo et al., 2010) and these genes are mostly involved in the p53 signaling, apoptosis, and cancer pathways when analyzed with the DAVID program (Huang et al., 2007).

2.2 Reprogramming Process-Dependent-Induced Genes

The regulatory status of induced genes can be affected by reprogramming transcription factors, virus vector type, and culture conditions. Induced genes (278 up- and 128 downregulated) most likely appear through the binding of ectopically expressed reprogramming transcription factors Oct4, Sox2, KLF4, and c-Myc (Takahashi et al., 2007; Nakatake et al., 2006). The auxiliary role of the other two factors LIN28 and Nanog

(Pan and Thomson, 2007) is also reported. In order to understand the activities of induced genes, transcription regulation analysis has been applied to this category.

2.3 iPSC-Reprogramming Factor Motifs

Because no binding matrices for the above reprogramming factors were available for *Homo sapiens* (humans), we have constructed matrices from experimental immunoprecipitation data (Boyer et al., 2005) listing 1500 genes with approximate binding locations within the range of 100–300 nucleotides for the Oct3/4, Sox2, and Nanog reprogramming factors.

Human position-specific scoring matrices (PSSM) for OCT4, SOX2, and NANOG were constructed from the regulatory regions of experimentally verified promoters (Boyer et al., 2012) using the MEME motif discovery tool, as depicted in Fig. 2. This step was taken because human PSSM for those transcription factors does not exist in any database. For example, the following genes were used for the matrices construction: OCT4 matrix from NANOG, OSR2, MSC, KCNN2, PCTK2, and RORB; SOX2 matrix from GREG2, SCN3A, NELL1, THBS2, HIST1H4D, and INHBA; and NANOG matrix from ONECUT1, GSC, PRKCDBP, FOXB1, and FGFR2. As a result, five motif matrices have been constructed with Meta-MEME: OCT4 [type1], OCT4 [type2], OCT3/4_SOX2, SOX2, and NANOG, as shown in Fig. 2. KLF4 and cMYC matrices were obtained from the JASPAR (Sandelin et al., 2004) and Transfac32 public databases. The LIN28 (Peng et al., 2010) matrix was acquired from publication. Eight motif matrices are provided in Supplementary Information on https://doi.org/10.1016/B978-0-12-809556-0.00013-7.

2.4 Promoter Analysis of Reprogramming Induced Genes

Regulatory analysis of 406 induced genes was performed using the above described matrices with the ExPlain3.0 suit (Wingender et al., 1996). Promoter regions to 5000 base pairs upstream and 500 base pairs downstream from the transcriptional start site (TSS) of the integrated TRANSPro database (Wingender et al., 1996) were used.

The above-mentioned matrices were then combined into one "*Master_gene*" profile and used in F-match software within the ExPlain3.0 suit. F-match evaluates the set of promoters and for each matrix tries to find two thresholds: one, *th*-max, which provides the maximum ratio between the frequency of matches in the promoter of in focus (control set, "C") and background promoters (background, "B") (overrepresented sites), and the second threshold, *th*-min, which minimizes the same ratio (underrepresented sites). A binominal distribution of the sites between each control promoter dataset and respective background dataset is calculated and the *P*-value is assigned. To test if each up- and down-regulated gene dataset identified in each analyzed cell line have overdistributed transcription factors, background datasets of the same size selected randomly from a

Ab initio motif finding by MEME program

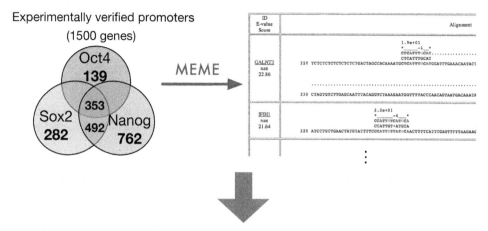

Motif constructed with Meta-Meme (Hidden Markov Model)

No.	Motif	Consensus*	L
1	Oct4[type1]	AKGMAAMKGAG	11
2	Oct4[type2]	TTGCATT	8
3	Oct3/4_Sox2	WGACATDACAAWGG	13
4	Sox2	WACAAWGS	8
5	Nanog	TAATKK	6
6	Klf4	WRRRRARRG	10
7	c-Myc	CACGTGS	7
8	LIN28	GCCACACCAYCGC	13

HMM Graph Representation

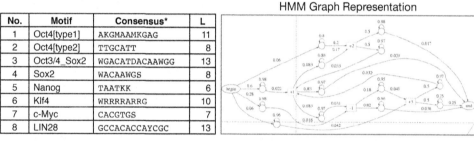

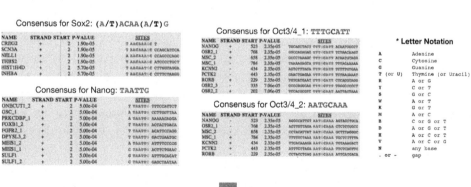

Analysis of Reprogramming Induced genes (406)

Fig. 2 Analytic workflow for creation of iPSC reprogramming factor binding matrices.

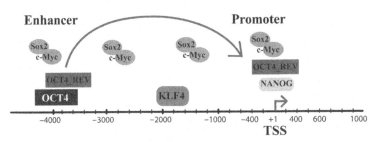

Fig. 3 iPSC reprogramming factor binding model based on allocations of predicted transcription factor binding sites by the F-match tool in ExPlain3.0 suite. TSS designates a transcriptional start site.

set of housekeeping genes (997 genes in were created). Overrepresented consensus motifs that appeared in three housekeeping background datasets were used as background motifs to calculate the *P*-value for the respective motifs in the control set using F-match (Wingender et al., 1996).

The selection criteria for the promoter to be accepted as a "possessing predicted binding motif" was that the promoter should have at least two out of the three main transcription factor binding motifs OCT4, SOX2, and NANOG under a *P*-value <0.001. LIN28 was not identified in the promoters of induced genes. As a result, an iPSC reprogramming factor binding model has been constructed, as shown in Fig. 3. NANOG was identified in the promoters of upregulated groups of 13 cell lines and in downregulated groups of seven cell lines (Polouliakh, 2013). The fact that NANOG was not one of the reprogramming factors implies the possibility of its ectopic activation in the course of reprogramming (Jiang et al., 2011). Fig. 3 depicts a promoter model with the approximated transcription factor binding allocation identified in our study. OCT4 binds between −4800 (SD ± 200) and −3600 (SD ± 200) ("−" refers to minus strand and "+" refers to plus strand), The OCT4_rev-complement binds between −3700 (SD ± 1000) and −2700 (SD ± 800) (first OCT4 position) and between −400 (SD ± 100) and +500 (SD ± 0.0) (second OCT4 position). KLF4 binds between −2000 (SD ± 600) and −1600 (SD ± 700), and NANOG constantly binds between −400 (SD ± 100) and +400 (SD ± 100) position to the TSS. SOX2, and c-Myc positions were not related to TSS. The reverse-complementary predicted sites of OCT4 were found in all cell lines.

2.5 Transcription Regulation Analysis Revealed Four Groups of Genes

Transcription regulation analysis revealed the existence of four groups of genes. The first group is genes with transcription motifs but without a significant hit to the particular GO category over a group. This group is termed the "master" gene group, and there are 42 upregulated and 25 downregulated genes shared by more than two cell lines. The second group is genes with a significant hit (*P*-value ≤0.05) to the "development" GO term and transcription factor motifs, and it is named the "master_development" gene group.

It has 36 upregulated genes and eight downregulated genes shared by more than two cell lines. The third group is genes with a significant hit to the GO "development" category and without predicted transcription factor binding motifs upon our selection criteria. We call it the "development" gene group, and it has 89 upregulated genes and 44 downregulated genes found in more than two cell lines. The fourth group is genes without a significant hit to any particular GO term over a group and without predicted transcription factor binding motifs. We call this group "others" and obtained 111 upregulated genes and 51 downregulated genes for this group (Polouliakh, 2013). Noteworthy is that 18% (SD ± 11.01) of upregulated genes and 23% (SD ± 9.2) of downregulated genes in each cell line of the induced genes category belong to the "master" and "master_development" groups, that is, the proportion of such genes in each cell line is similar in size.

The following pathways were identified in up- and downregulated or either group in the induced genes category using the DAVID annotation tool (Huang et al., 2007): the calcium signaling pathway (4.00E − 03, up-, down-), the cell adhesion molecules (CAMs) pathway (1.27E − 02, up-), the PPAR signaling pathway (1.83E − 02, up-), and the tight junction (2.69E − 02, down-) and neuroactive ligand-receptor interaction (3.45E − 02, down-) pathways. Calcium-related genes in both up- and downregulated groups might imply the possibility for induction of differentiation and development.

3. COMPARATIVE GENOMIC ANALYSIS iPSC TRANSCRIPTION REGULATION WITH SHOE

iPSC reprogramming factor matrices were incorporated into the original comparative genomic analysis tool SHOE (Sequence HOmology in higher Eukaryotes) (Yoshino et al., 2016; Polouliakh et al., 2006) to facilitate investigations of iPSC reprogramming motif evolutionary conservation between three species: human, mouse, and rat. This kind of investigation can fortify experimental analysis because the cooccurrence of the same motif in the promoters of orthologous species increases motif likelihood in humans. Fig. 4 demonstrates that 6 out of 42 iPSC reprogramming factor induced genes have promoters conserved between human, mouse, and rat. SHOE is a standalone software with demo allocated on http://ec2-54-150-223-65.ap-northeast-1.compute. amazonaws.com site.

4. CONCLUSIONS

Reprogramming-resistant genes are classified into two categories of induced genes and inherited genes, depending on their expression status in somatic cells of origin. Induced genes exhibit bivalent (H3K4K27) modification status in ES cells with the predominance of intermediate and low CpG density promoters and promoters with nondefined CpG density ("ND"). On the contrary, the inherited genes category was enriched in univalent (H3K4) modification status in ES cells and showed a preponderance for high CpG density promoter genes (Polouliakh, 2013).

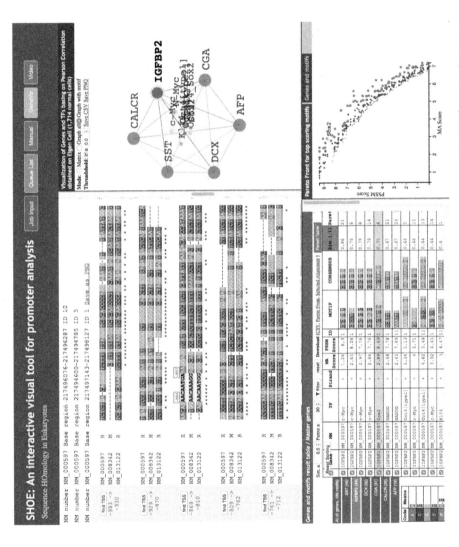

Fig. 4 Multiple alignment on human-mouse-rat on six reprogramming-resistant master genes with iPSC factor in their promoters. IGFBP2 gene is highlighted in the result table and in the network. The SOX2 motif is selected in the results table as significantly conserved across three species, and its location to TSS is shown in the alignment.

Regarding the characteristics of reprogramming-resistant genes on the pathway level, the inherited gene category included cancer and apoptosis-related pathways such as the focal adhesion and the p53 signaling pathway, which were also observed in the recent results. This may affect the unwanted tumorigenic propensity of iPS cells, and further experimental verification of this issue is required. The induced gene category was enriched in the calcium-signaling pathway, cell adhesion, PPAR signaling, and the tight junction. These pathways may contribute to the embryogenesis, development, and immune response, but the biological implication of such differential expressions is yet to be elucidated.

Transcription regulation analysis on iPSC induced genes helped to discover four subgroups within this category of genes: master (with iPSC factor in promoters, various GO annotation), master development (with iPSC factor in promoters, development GO function), development (no iPSC factor in promoters, development GO annotation), and others (no iPSC factors in promoters, various GO annotation). Five novel human iPSC factor matrices for *Homo sapiens* that are OCT4 [type 1], OCT4 [type 2], OCT3/4_SOX2, SOX2, and NANOG were created in this analysis, in addition to those already known, using experimental data through the motif discovery MEME tool. They are provided for users with already known matrices for KLF4, c-MYC, and LIN-28. The iPSC reprogramming factor binding model for the induced genes category was identified. Further, incorporation of iPSC factor matrices in the promoter analysis tool SHOE helps to extend the analytical workflow on a cross-species level for iPSC and other cells.

REFERENCES

Bailey, T.L., Elkan, C., 1995. The value of prior knowledge in discovering motifs with MEME. Proc. Int. Conf. Intell. Syst. Mol. Biol. 3, 21–29.

Bailey, T.L., Johnson, J., Grant, C.E., Noble, W.S., 2015. The MEME suite. Nucleic Acids Res. 43, W39–W49.

Berezikov, E., Guryev, V., Cuppen, E., 2005. CONREAL web server: identification and visualization of conserved transcription factor binding sites. Nucleic Acids Res. 33, W447–W450.

Boyer, L.A., Lee, T.I., Cole, M.F., Johnstone, S.E., Levine, S.S., Zucker, J.P., Guenther, M.G., Kumar, R.M., Murray, H.L., Jenner, R.G., Gifford, D.K., Melton, D.A., Jaenisch, R., Young, R.A., 2005. Core transcriptional regulatory circuitry in human embryonic stem cells. Cell 122, 947–956.

Boyer, J.Z., Jandova, J., Janda, J., Vleugels, F.R., Elliott, D.A., Sligh, J.E., 2012. Resveratrol-sensitized UVA induced apoptosis in human keratinocytes through mitochondrial oxidative stress and pore opening. J. Photochem. Photobiol. B 113, 42–50.

Chin, M.H., Mason, M.J., Xie, W., Volinia, S., Singer, M., Peterson, C., Ambartsumyan, G., Aimiuwu, O., Richter, L., Zhang, J., Khvorostov, I., Ott, V., Grunstein, M., Lavon, N., Benvenisty, N., Croce, C.M., Clark, A.T., Baxter, T., Pyle, A.D., Teitell, M.A., Pelegrini, M., Plath, K., Lowry, W.E., 2009. Induced pluripotent stem cells and embryonic stem cells are distinguished by gene expression signatures. Cell Stem Cell 5, 111–123.

Eckhardt, F., Lewin, J., Cortese, R., Rakyan, V.K., Attwood, J., Burger, M., Burton, J., Cox, T.V., Davies, R., Down, T.A., Haefliger, C., Horton, R., Howe, K., Jackson, D.K., Kunde, J., Koenig, C., Liddle, J., Niblett, D., Otto, T., Pettett, R., Seemann, S., Thompson, C., West, T.,

Rogers, J., Olek, A., Berlin, K., Beck, S., 2006. DNA methylation profiling of human chromosomes 6, 20 and 22. Nat. Genet. 38, 1378–1385.

Fang, F., Blanchette, M., 2006. FootPrinter3: phylogenetic footprinting in partially alignable sequences. Nucleic Acids Res. 34, W617–W620.

Ghosh, S., Matsuoka, Y., Asai, Y., Hsin, K.Y., Kitano, H., 2011. Software for systems biology: from tools to integrated platforms. Nat. Rev. Genet. 12, 821–832.

Gupta, M.K., Illich, D.J., Gaarz, A., Matzkies, M., Nguemo, F., Pfannkuche, K., Liang, H., Classen, S., Reppel, M., Schultze, J.L., Hescheler, J., Saric, T., 2010. Global transcriptional profiles of beating clusters derived from human induced pluripotent stem cells and embryonic stem cells are highly similar. BMC Dev. Biol. 10, 98.

Hertz, G.Z., Stormo, G.D., 1999. Identifying DNA and protein patterns with statistically significant alignments of multiple sequences. Bioinformatics 15, 563–577.

Huang, D.W., Sherman, B.T., Tan, Q., Kir, J., Liu, D., Bryant, D., Guo, Y., Stephens, R., Baseler, M.W., Lane, H.C., Lempicki, R.A., 2007. DAVID Bioinformatics Resources: expanded annotation database and novel algorithms to better extract biology from large gene lists. Nucleic Acids Res. 35, W169–W175.

Jablonska, A., Polouliakh, N., 2014. In silico discovery of novel transcription factors regulated by mTOR-pathway activities. Front. Cell Dev. Biol. 2, 23.

Jiang, J., Ding, G., Lin, J., Zhang, M., Shi, L., Lv, W., Yang, H., Xiao, H., Pei, G., Li, Y., Wu, J., Li, J., 2011. Different developmental potential of pluripotent stem cells generated by different reprogramming strategies. J. Mol. Cell Biol. 3, 197–199.

Keich, U., Pevzner, P.A., 2002. Finding motifs in the twilight zone. Bioinformatics 18, 1374–1381.

Lawrence, C.E., Altschul, S.F., Boguski, M.S., Liu, J.S., Neuwald, A.F., Wootton, J.C., 1993. Detecting subtle sequence signals: a Gibbs sampling strategy for multiple alignment. Science 262, 208–214.

Lee, T.Y., Chang, W.C., Hsu, J.B., Chang, T.H., Shien, D.M., 2012. GPMiner: an integrated system for mining combinatorial cis-regulatory elements in mammalian gene group. BMC Genomics 13 (Suppl. 1), S3.

Liu, X., Brutlag, D.L., Liu, J.S., 2001. BioProspector: discovering conserved DNA motifs in upstream regulatory regions of co-expressed genes. Pac. Symp. Biocomput., 127–138.

Meissner, A., Mikkelsen, T.S., Gu, H., Wernig, M., Hanna, J., Sivachenko, A., Zhang, X., Bernstein, B.E., Nusbaum, C., Jaffe, D.B., Gnirke, A., Jaenisch, R., Lander, E.S., 2008. Genome-scale DNA methylation maps of pluripotent and differentiated cells. Nature 454, 766–770.

Moses, A.M., Chiang, D.Y., Kellis, M., Lander, E.S., Eisen, M.B., 2003. Position specific variation in the rate of evolution in transcription factor binding sites. BMC Evol. Biol. 3, 19.

Muller, F.J., Schuldt, B.M., Williams, R., Mason, D., Altun, G., Papapetrou, E.P., Danner, S., Goldmann, J.E., Herbst, A., Schmidt, N.O., Aldenhoff, J.B., Laurent, L.C., Loring, J.F., 2011. A bioinformatic assay for pluripotency in human cells. Nat. Methods 8, 315–317.

Nakatake, Y., Fukui, N., Iwamatsu, Y., Masui, S., Takahashi, K., Yagi, R., Yagi, K., Miyazaki, J., Matoba, R., Ko, M.S., Niwa, H., 2006. Klf4 cooperates with Oct3/4 and Sox2 to activate the Lefty1 core promoter in embryonic stem cells. Mol. Cell. Biol. 26, 7772–7782.

Pan, G., Thomson, J.A., 2007. Nanog and transcriptional networks in embryonic stem cell pluripotency. Cell Res. 17, 42–49.

Peng, S., Maihle, N.J., Huang, Y., 2010. Pluripotency factors Lin28 and Oct4 identify a sub-population of stem cell-like cells in ovarian cancer. Oncogene 29, 2153–2159.

Polo, J.M., Liu, S., Figueroa, M.E., Kulalert, W., Eminli, S., Tan, K.Y., Apostolou, E., Stadtfeld, M., Li, Y., Shioda, T., Natesan, S., Wagers, A.J., Melnick, A., Evans, T., Hochedlinger, K., 2010. Cell type of origin influences the molecular and functional properties of mouse induced pluripotent stem cells. Nat. Biotechnol. 28, 848–855.

Polouliakh, N., 2013. Reprogramming resistant genes: in-depth comparison of gene expressions among iPS, ES, and somatic cells. Front. Physiol. 4, 7.

Polouliakh, N., Konno, M., Horton, P., Nakai, K., 2004. Parameter landscape analysis for common motif discovery programs. In: Regulatory Genomics. Lecture Notes in Computer Science, vol. 3318. Springer, pp. 79–81.

Polouliakh, N., Natsume, T., Harada, H., Fujibuchi, W., Horton, P., 2006. Comparative genomic analysis of transcription regulation elements involved in human map kinase G-protein coupling pathway. J. Bioinforma. Comput. Biol. 4, 469–482.

Poluliakh, N., Takagi, T., Nakai, K., 2003. Melina: motif extraction from promoter regions of potentially co-regulated genes. Bioinformatics 19, 423–424.

Sandelin, A., Alkema, W., Engstrom, P., Wasserman, W.W., Lenhard, B., 2004. JASPAR: an open-access database for eukaryotic transcription factor binding profiles. Nucleic Acids Res. 32, D91–D94.

Serizawa, S., Miyamichi, K., Nakatani, H., Suzuki, M., Saito, M., Yoshihara, Y., Sakano, H., 2003. Negative feedback regulation ensures the one receptor-one olfactory neuron rule in mouse. Science 302, 2088–2094.

Sinha, S., Tompa, M., 2000. A statistical method for finding transcription factor binding sites. Proc. Int. Conf. Intell. Syst. Mol. Biol. 8, 344–354.

Stormo, G.D., Hartzell 3rd., G.W., 1989. Identifying protein-binding sites from unaligned DNA fragments. Proc. Natl. Acad. Sci. U. S. A. 86, 1183–1187.

Takahashi, K., Tanabe, K., Ohnuki, M., Narita, M., Ichisaka, T., Tomoda, K., Yamanaka, S., 2007. Induction of pluripotent stem cells from adult human fibroblasts by defined factors. Cell 131, 861–872.

Wingender, E., Dietze, P., Karas, H., Knuppel, R., 1996. TRANSFAC: a database on transcription factors and their DNA binding sites. Nucleic Acids Res. 24, 238–241.

Wolfertstetter, F., Frech, K., Herrmann, G., Werner, T., 1996. Identification of functional elements in unaligned nucleic acid sequences by a novel tuple search algorithm. Comput. Appl. Biosci. 12, 71–80.

Yoshino, A., Polouliakh, N., Meguro, A., Takeuchi, M., Kawagoe, T., Mizuki, N., 2016. Chum salmon egg extracts induce upregulation of collagen type I and exert antioxidative effects on human dermal fibroblast cultures. Clin. Interv. Aging 11, 1159–1168.

INDEX

Note: Page numbers followed by *f* indicate figures and *t* indicate tables.